ROBUST ADAPTIVE DYNAMIC PROGRAMMING

ROBUST ADAPTIVE DYNAMIC PROGRAMMING

YU JIANG
The MathWorks, Inc.

ZHONG-PING JIANG
New York University

WILEY

Published by John Wiley & Sons, Inc., Hoboken, New Jersey.
Published simultaneously in Canada.

For general information on our other products and services or for technical support, please contact our Customer Care Department within the United States at (800) 762-2974, outside the United States at (317) 572-3993 or fax (317) 572-4002.

Wiley also publishes its books in a variety of electronic formats. Some content that appears in print may not be available in electronic formats. For more information about Wiley products, visit our web site at www.wiley.com.

Library of Congress Cataloging-in-Publication Data is available.

ISBN: 978-1-119-13264-6

Printed in the United States of America.

10 9 8 7 6 5 4 3 2 1

To my mother, Misi, and Xiaofeng
—Yu Jiang

To my family
—Zhong-Ping Jiang

CONTENTS

ABOUT THE AUTHORS

Yu Jiang is a Software Engineer with the Control Systems Toolbox Team at The MathWorks, Inc. He received a B.Sc. degree in Applied Mathematics from Sun Yat-sen University, Guangzhou, China, a M.Sc. degree in Automation Science and Engineering from South China University of Technology, Guangzhou, China, and a Ph.D. degree in Electrical Engineering from New York University. His research interests include adaptive dynamic programming and other numerical methods in control and optimization. He was the recipient of the Shimemura Young Author Prize (with Prof. Z.P. Jiang) at the 9th Asian Control Conference in Istanbul, Turkey, 2013.

Zhong-Ping Jiang is a Professor of Electrical and Computer Engineering at the Tandon School of Engineering, New York University. His main research interests include stability theory, robust/adaptive/distributed nonlinear control, adaptive dynamic programming and their applications to information, mechanical and biological systems. In these areas, he has written 3 books, 14 book chapters and is the (co-)author of over 182 journal papers and numerous conference papers. His work has received 15,800 citations with an h-index of 63 according to Google Scholar. Professor Jiang is a Deputy co-Editor-in-Chief of the *Journal of Control and Decision*, a Senior Editor for the IEEE Transactions on Control Systems Letters, and has served as an editor, a guest editor and an associate editor for several journals in Systems and Control. Prof. Jiang is a Fellow of the IEEE and a Fellow of the IFAC.

PREFACE AND ACKNOWLEDGMENTS

This book covers the topic of adaptive optimal control (AOC) for continuous-time systems. An adaptive optimal controller can gradually modify itself to adapt to the controlled system, and the adaptation is measured by some performance index of the closed-loop system. The study of AOC can be traced back to the 1970s, when researchers at the Los Alamos Scientific Laboratory (LASL) started to investigate the use of adaptive and optimal control techniques in buildings with solar-based temperature control. Compared with conventional adaptive control, AOC has the important ability to improve energy conservation and system performance. However, even though there are various ways in AOC to compute the optimal controller, most of the previously known approaches are model-based, in the sense that a model with a fixed structure is assumed before designing the controller. In addition, these approaches do not generalize to nonlinear models.

On the other hand, quite a few model-free, data-driven approaches for AOC have emerged in recent years. In particular, adaptive/approximate dynamic programming (ADP) is a powerful methodology that integrates the idea of reinforcement learning (RL) observed from mammalian brain with decision theory so that controllers for man-made systems can learn to achieve optimal performance in spite of uncertainty about the environment and the lack of detailed system models. Since the 1960s, RL has been brought to the computer science and control science literature as a way to study artificial intelligence, and has been successfully applied to many discrete-time systems, or Markov Decision Processes (MDPs). However, it has always been challenging to generalize those results to the controller design of physical systems. This is mainly because the state space of a physical control system is generally continuous and unbounded, and the states are continuous in time. Therefore, the convergence and the stability properties have to be carefully studied for ADP-based approaches. The main purpose of this book is to introduce the recently developed framework, known as robust adaptive dynamic programming (RADP), for data-driven, non-model based adaptive optimal control design for both linear and nonlinear continuous-time systems.

In addition, this book is intended to address in a systematic way the presence of dynamic uncertainty. Dynamic uncertainty exists ubiquitously in control engineering. It is primarily caused by the dynamics which are part of the physical system but are either difficult to be mathematically modeled or ignored for the sake of controller design and system analysis. Without addressing the dynamic uncertainty, controller designs based on the simplified model will most likely fail when applied to the physical system. In most of the previously developed ADP or other RL methods, it is assumed that the full-state information is always available, and therefore the system order must be known. Although this assumption excludes the existence of any dynamic uncertainty, it is apparently too strong to be realistic. For a physical model on a relatively large scale, knowing the exact number of state variables can be difficult, not to mention that not all state variables can be measured precisely. For example, consider a power grid with a main generator controlled by the utility company and small distributed generators (DGs) installed by customers. The utility company should not neglect the dynamics of the DGs, but should treat them as dynamic uncertainties when controlling the grid, such that stability, performance, and power security can be always maintained as expected.

The book is organized in four parts. First, an overview of RL, ADP, and RADP is contained in Chapter 1. Second, a few recently developed continuous-time ADP methods are introduced in Chapters 2, 3, and 4. Chapter 2 covers the topic of ADP for uncertain linear systems. Chapters 3 and 4 provide neural network-based and sum-of-squares (SOS)-based ADP methodologies to achieve semi-global and global stabilization for uncertain nonlinear continuous-time systems, respectively. Third, Chapters 5 and 6 focus on RADP for linear and nonlinear systems, with dynamic uncertainties rigorously addressed. In Chapter 5, different robustification schemes are introduced to achieve RADP. Chapter 6 further extends the RADP framework for large-scale systems and illustrates its applicability to industrial power systems. Finally, Chapter 7 applies ADP and RADP to study the sensorimotor control of humans, and the results suggest that humans may be using very similar approaches to learn to coordinate movements to handle uncertainties in our daily lives.

This book makes a major departure from most existing texts covering the same topics by providing many practical examples such as power systems and human sensorimotor control systems to illustrate the effectiveness of our results. The book uses MATLAB in each chapter to conduct numerical simulations. MATLAB is used as a computational tool, a programming tool, and a graphical tool. Simulink, a graphical programming environment for modeling, simulating, and analyzing multidomain dynamic systems, is used in Chapter 2. The third-party MATLAB-based software SOSTOOLS and CVX are used in Chapters 4 and 5 to solve SOS programs and semidefinite programs (SDP). All MATLAB programs and the Simulink model developed in this book as well as extension of these programs are available at http://yu-jiang.github.io/radpbook/

The development of this book would not have been possible without the support and help of many people. The authors wish to thank Prof. Frank Lewis and Dr. Paul Werbos whose seminal work on adaptive/approximate dynamic programming has laid down the foundation of the book. The first-named author (YJ) would like to thank his Master's Thesis adviser Prof. Jie Huang for guiding him into the area of nonlinear control, and Dr. Yebin Wang for offering him a summer research internship position at Mitsubishi Electric Research Laboratories, where parts of the ideas in Chapters 4 and 5 were originally inspired. The second-named author (ZPJ) would like to acknowledge his colleagues—specially Drs. Alessandro Astolfi, Lei Guo, Iven Mareels, and Frank Lewis—for many useful comments and constructive criticism on some of the research summarized in the book. He is grateful to his students for the boldness in entering the interesting yet still unpopular field of data-driven adaptive optimal control. The authors wish to thank the editors and editorial staff, in particular, Mengchu Zhou, Mary Hatcher, Brady Chin, Suresh Srinivasan, and Divya Narayanan, for their efforts in publishing the book. We thank Tao Bian and Weinan Gao for collaboration on generalizations and applications of ADP based on the framework of RADP presented in this book. Finally, we thank our families for their sacrifice in adapting to our hard-to-predict working schedules that often involve dynamic uncertainties. From our family members, we have learned the importance of exploration noise in achieving the desired trade-off between robustness and optimality. The bulk of this research was accomplished while the first-named author was working toward his Ph.D. degree in the Control and Networks Lab at New York University Tandon School of Engineering. The authors wish to acknowledge the research funding support by the National Science Foundation.

YU JIANG
Wellesley, Massachusetts

ZHONG-PING JIANG
Brooklyn, New York

ACRONYMS

ADP	Adaptive/approximate dynamic programming
AOC	Adaptive optimal control
ARE	Algebraic Riccati equation
DF	Divergent force field
DG	Distributed generator/generation
DP	Dynamic programming
GAS	Global asymptotic stability
HJB	Hamilton-Jacobi-Bellman (equation)
IOS	Input-to-output stability
ISS	Input-to-state stability
LQR	Linear quadratic regulator
MDP	Markov decision process
NF	Null-field
PE	Persistent excitation
PI	Policy iteration
RADP	Robust adaptive dynamic programming
RL	Reinforcement learning
SDP	Semidefinite programming
SOS	Sum-of-squares
SUO	Strong unboundedness observability
VF	Velocity-dependent force field
VI	Value iteration

GLOSSARY

$\lvert \cdot \rvert$	The Euclidean norm for vectors, or the induced matrix norm for matrices
$\lVert \cdot \rVert$	For any piecewise continuous function $u : \mathbb{R}_+ \to \mathbb{R}^m$, $\lVert u \rVert = \sup\{\lvert u(t) \rvert, t \geq 0\}$
$\otimes$	Kronecker product
C^1	The set of all continuously differentiable functions
$J_D^{\oplus}$	The cost for the coupled large-scale system
$J_D^{\odot}$	The cost for the decoupled large-scale system
$\mathcal{P}$	The set of all functions in C^1 that are also positive definite and radially unbounded
$\mathcal{L}(\cdot)$	Infinitesimal generator
$\mathbb{R}$	The set of all real numbers
$\mathbb{R}_+$	The set of all non-negative real numbers
$\mathbb{R}[x]_{d_1,d_2}$	The set of all polynomials in $x \in \mathbb{R}^n$ with degree no less than $d_1 > 0$ and no greater than d_2
$\mathrm{vec}(\cdot)$	$\mathrm{vec}(A)$ is defined to be the mn-vector formed by stacking the columns of $A \in \mathbb{R}^{n \times m}$ on top of another, that is, $\mathrm{vec}(A) = [a_1^T a_2^T \cdots a_m^T]^T$, where $a_i \in \mathbb{R}^n$, with $i = 1, 2, \ldots, m$, are the columns of A
$\mathbb{Z}_+$	The set of all non-negative integers
$[x]_{d_1,d_2}$	The vector of all $\binom{n+d_2}{d_2} - \binom{n+d_1-1}{d_1-1}$ distinct monic monomials in $x \in \mathbb{R}^n$ with degree no less than $d_1 > 0$ and no greater than d_2
∇	∇V refers to the gradient of a differentiable function $V : \mathbb{R}^n \to \mathbb{R}$

CHAPTER 1

INTRODUCTION

1.1 FROM RL TO RADP

1.1.1 Introduction to RL

Reinforcement learning (RL) is originally observed from the learning behavior in humans and other mammals. The definition of RL varies in different literature. Indeed, learning a certain task through trial-and-error can be considered as an example of RL. In general, an RL problem requires the existence of an *agent*, that can interact with some unknown *environment* by taking *actions*, and receiving a *reward* from it. Sutton and Barto referred to RL as *how to map situations to actions so as to maximize a numerical reward signal* [47]. Apparently, maximizing a reward is equivalent to minimizing a *cost*, which is used more frequently in the context of optimal control [32]. In this book, a mapping between situations and actions is called a *policy*, and the goal of RL is to learn an optimal policy such that a predefined cost is minimized.

As a unique learning approach, RL does not require a supervisor to teach an agent to take the optimal action. Instead, it focuses on how the agent, through interactions with the unknown environment, should modify its own actions toward the optimal one (Figure 1.1). An RL iteration generally contains two major steps. First, the agent evaluates the cost under the current policy, through interacting with the environment. This step is known as *policy evaluation*. Second, based on the evaluated cost, the agent adopts a new policy aiming at further reducing the cost. This is the step of *policy improvement*.

Robust Adaptive Dynamic Programming, First Edition. Yu Jiang and Zhong-Ping Jiang.
© 2017 by The Institute of Electrical and Electronics Engineers, Inc. Published 2017 by John Wiley & Sons, Inc.

FIGURE 1.1   Illustration of RL. The agent takes an action to interact with the unknown environment, and evaluates the resulting cost, based on which the agent can further improve the action to reduce the cost.

As an important branch in machine learning theory, RL has been brought to the computer science and control science literature as a way to study artificial intelligence in the 1960s [37, 38, 54]. Since then, numerous contributions to RL, from a control perspective, have been made (see, e.g., [2, 29, 33, 34, 46, 53, 56]). Recently, AlphaGo, a computer program developed by Google DeepMind, is able to improve itself through reinforcement learning and has beaten professional human Go players [44]. It is believed that significant attention will continuously be paid to the study of reinforcement learning, since it is a promising tool for us to better understand the true intelligence in human brains.

1.1.2 Introduction to DP

On the other hand, dynamic programming (DP) [4] offers a theoretical way to solve multistage decision-making problems. However, it suffers from the inherent computational complexity, also known as the *curse of dimensionality* [41]. Therefore, the need for approximative methods has been recognized as early as in the late 1950s [3]. In [15], an iterative technique called policy iteration (PI) was devised by Howard for Markov decision processes (MDPs). Also, Howard referred to the iterative method developed by Bellman [3, 4] as value iteration (VI). Computing the optimal solution through successive approximations, PI is closely related to learning methods. In 1968, Werbos pointed out that PI can be employed to perform RL [58]. Starting from then, many real-time RL methods for finding online optimal control policies have emerged and they are broadly called approximate/adaptive dynamic programming (ADP) [31, 33, 41, 43, 55, 60–65, 68], or neurodynamic programming [5]. The main feature of ADP [59, 61] is that it employs ideas from RL to achieve online approximation of the value function, without using the knowledge of the system dynamics.

1.1.3 The Development of ADP

The development of ADP theory consists of three phases. In the first phase, ADP was extensively investigated within the communities of computer science and operations research. PI and VI are usually employed as two basic algorithms. In [46], Sutton introduced the temporal difference method. In 1989, Watkins proposed the well-known Q-learning method in his PhD thesis [56]. Q-learning shares similar features with the action-dependent heuristic dynamic programming (ADHDP) scheme proposed by Werbos in [62]. Other related research work under a discrete time and discrete state-space Markov decision process framework can be found in [5, 6, 8, 9, 41, 42, 48, 47] and references therein. In the second phase, stability is brought into the context of ADP while real-time control problems are studied for dynamic systems. To the best of our knowledge, Lewis and his co-workers are the first who contributed to the integration of stability theory and ADP theory [33]. An essential advantage of ADP theory is that an optimal control policy can be obtained via a recursive numerical algorithm using online information without solving the Hamilton-Jacobi-Bellman (HJB) equation (for nonlinear systems) and the algebraic Riccati equation (ARE) (for linear systems), even when the system dynamics are not precisely known. Related optimal feedback control designs for linear and nonlinear dynamic systems have been proposed by several researchers over the past few years; see, for example, [7, 10, 39, 40, 50, 52, 66, 69]. While most of the previous work on ADP theory was devoted to discrete-time (DT) systems (see [31] and references therein), there has been relatively less research for the continuous-time (CT) counterpart. This is mainly because ADP is considerably more difficult for CT systems than for DT systems. Indeed, many results developed for DT systems [35] cannot be extended straightforwardly to CT systems. As a result, early attempts were made to apply Q-learning for CT systems via discretization technique [1, 11]. However, the convergence and stability analysis of these schemes are challenging. In [40], Murray et. al proposed an implementation method which requires the measurements of the derivatives of the state variables. As said previously, Lewis and his co-workers proposed the first solution to stability analysis and convergence proofs for ADP-based control systems by means of linear quadratic regulator (LQR) theory [52]. A synchronous policy iteration scheme was also presented in [49]. For CT linear systems, the partial knowledge of the system dynamics (i.e., the input matrix) must be precisely known. This restriction has been completely removed in [18]. A nonlinear variant of this method can be found in [22] and [23].

The third phase in the development of ADP theory is related to extensions of previous ADP results to nonlinear uncertain systems. Neural networks and game theory are utilized to address the presence of uncertainty and nonlinearity in control systems. See, for example, [14, 31, 50, 51, 57, 67, 69, 70]. An implicit assumption in these papers is that the system order is known and that the uncertainty is static, not dynamic. The presence of dynamic uncertainty has not been systematically addressed in the literature of ADP. By dynamic uncertainty, we refer to the mismatch between the nominal model (also referred to as the *reduced-order system*) and the real plant when the order of the nominal model is lower than the order of the real system. A closely related topic of research is how to account for the effect of unseen variables [60]. It is quite common that the full-state information is often missing in many engineering applications and only the output measurement or partial-state measurements are available. Adaptation of the existing ADP theory to this practical scenario is important yet non-trivial. Neural networks are sought for addressing the state estimation problem

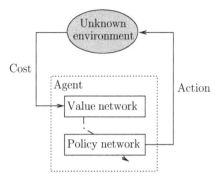

FIGURE 1.2 Illustration of the ADP scheme.

[12, 28]. However, the stability analysis of the estimator/controller augmented system is by no means easy, because the total system is highly interconnected and often strongly nonlinear. The configuration of a standard ADP-based control system is shown in Figure 1.2.

Our recent work [17, 19, 20, 21] on the development of robust ADP (for short, RADP) theory is exactly targeted at addressing these challenges.

1.1.4 What Is RADP?

RADP is developed to address the presence of dynamic uncertainty in linear and nonlinear dynamical systems. See Figure 1.3 for an illustration. There are several reasons for which we pursue a new framework for RADP. First and foremost, it is well known that building an exact mathematical model for physical systems often is

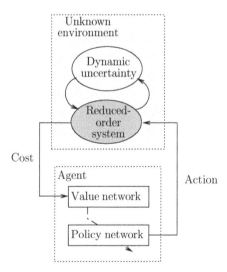

FIGURE 1.3 In the RADP learning scheme, a new component, known as dynamic uncertainty, is taken into consideration.

a hard task. Also, even if the exact mathematical model can be obtained for some particular engineering and biological applications, simplified nominal models are often more preferable for system analysis and control synthesis than the original complex system model. While we refer to the mismatch between the simplified nominal model and the original system as dynamic uncertainty here, the engineering literature often uses the term of *unmodeled dynamics* instead. Second, the observation errors may often be captured by dynamic uncertainty. From the literature of modern nonlinear control [25, 26, 30], it is known that the presence of dynamic uncertainty makes the feedback control problem extremely challenging in the context of nonlinear systems. In order to broaden the application scope of ADP theory in the presence of dynamic uncertainty, our strategy is to integrate tools from nonlinear control theory, such as Lyapunov designs, input-to-state stability theory [45], and nonlinear small-gain techniques [27]. This way RADP becomes applicable to wide classes of uncertain dynamic systems with incomplete state information and unknown system order/dynamics.

Additionally, RADP can be applied to large-scale dynamic systems as shown in our recent paper [20]. By integrating a simple version of the cyclic-small-gain theorem [36], asymptotic stability can be achieved by assigning appropriate weighting matrices for each subsystem. Further, certain suboptimality property can be obtained. Because of several emerging applications of practical importance such as smart electric grid, intelligent transportation systems, and groups of mobile autonomous agents, this topic deserves further investigations from an RADP point of view. The existence of unknown parameters and/or dynamic uncertainties and the limited information of state variables give rise to challenges for the decentralized or distributed controller design of large-scale systems.

1.2 SUMMARY OF EACH CHAPTER

This book is organized as follows. Chapter 2 studies ADP for uncertain linear systems, of which the only a priori knowledge is an initial, stabilizing static state-feedback control policy. Then, via policy iteration, the optimal control policy is approximated. Two ADP methods, on-policy learning and off-policy learning, are introduced to achieve online implementation of conventional policy iteration. As a result, the optimal control policy can be approximated using online measurements, instead of the knowledge of the system dynamics.

Chapter 3 further extends the ADP methods for uncertain affine nonlinear systems. To guarantee proper approximation of the value function and the control policy, neural networks are applied. Convergence and stability properties of the nonlinear ADP method are rigorously proved. It is shown that semi-global stabilization is attainable for a general class of continuous-time nonlinear systems, under the approximate optimal control policy.

Chapter 4 focuses on the theory of global adaptive dynamic programming (GADP). It aims at simultaneously improving the closed-loop system performance and achieving global asymptotic stability of the overall system at the origin. It is shown that the equality constraint used in policy evaluation can be relaxed to a sum-of-squares

(SOS) constraint. Hence, an SOS-based policy iteration is formulated, by relaxing the conventional policy iteration. In this new policy iteration algorithm, the control policy obtained at each iteration step is globally stabilizing. Similarly, the SOS-based policy iteration can be implemented online, without the need to identify the exact system dynamics.

Chapter 5 presents the new framework of RADP. In contrast to the ADP theory introduced in Chapters 2–4, RADP does not require all the state variables to be available, nor the system order assumed known. Instead, it incorporates a subsystem known as the dynamic uncertainty that interacts with a simplified reduced-order model. While ADP methods are performed on the reduced model, the interactions between the dynamic uncertainty and the simplified model are studied using tools borrowed from modern nonlinear system analysis and controller design. The learning objective in RADP is to achieve optimal performance of the reduced-order model in the absence of dynamic uncertainties, and maintain robustness of stability in the presence of the dynamic uncertainty.

Chapter 6 applies the RADP framework to solve the decentralized optimal control problem for a class of large-scale uncertain systems. In recent years, considerable attention has been paid to the stabilization of large-scale complex systems, as well as related consensus and synchronization problems. Examples of large-scale systems arise from ecosystems, transportation networks, and power systems. Often, in real-world applications, precise mathematical models are hard to build, and the model mismatch, caused by parametric and dynamic uncertainties, is thus unavoidable. This, together with the exchange of only local system information, makes the design problem challenging in the context of complex networks. In this chapter, the controller design for each subsystem only needs to utilize local state measurements without knowing the system dynamics. By integrating a simple version of the cyclic-small-gain theorem, asymptotic stability can be achieved by assigning appropriate nonlinear gains for each subsystem.

Chapter 7 studies sensorimotor control with static and dynamic uncertainties under the framework of RADP [18, 19, 21, 24]. The linear version of RADP is extended for stochastic systems by taking into account signal-dependent noise [13], and the proposed method is applied to study the sensorimotor control problem with both static and dynamic uncertainties. Results presented in this chapter suggest that the central nervous system (CNS) may use RADP-like learning strategy to coordinate movements and to achieve successful adaptation in the presence of static and/or dynamic uncertainties. In the absence of the dynamic uncertainties, the learning strategy reduces to an ADP-like mechanism.

All the numerical simulations in this book are developed using MATLAB® R2015a. Source code is available on the webpage of the book [16].

REFERENCES

[1] L. C. Baird. Reinforcement learning in continuous time: Advantage updating. In: Proceedings of the IEEE World Congress on Computational Intelligence, Vol. 4, pp. 2448–2453, Orlando, FL, 1994.

[2] A. G. Barto, R. S. Sutton, and C. W. Anderson. Neuronlike adaptive elements that can solve difficult learning control problems. *IEEE Transactions on Systems, Man and Cybernetics*, 13(5):834–846, 1983.

[3] R. Bellman and S. Dreyfus. Functional approximations and dynamic programming. *Mathematical Tables and Other Aids to Computation*, 13(68):247–251, 1959.

[4] R. E. Bellman. *Dynamic Programming*. Princeton University Press, Princeton, NJ, 1957.

[5] D. P. Bertsekas. *Dynamic Programming and Optimal Control*, 4th ed. Athena Scientific, Belmont, MA, 2007.

[6] D. P. Bertsekas and J. N. Tsitsiklis. *Neuro-Dynamic Programming*. Athena Scientific, Nashua, NH, 1996.

[7] S. Bhasin, N. Sharma, P. Patre, and W. Dixon. Asymptotic tracking by a reinforcement learning-based adaptive critic controller. *Journal of Control Theory and Applications*, 9(3):400–409, 2011.

[8] V. S. Borkar. *Stochastic Approximation: A Dynamical Systems Viewpoint*. Cambridge University Press, Cambridge, 2008.

[9] L. Busoniu, R. Babuska, B. De Schutter, and D. Ernst. *Reinforcement Learning and Dynamic Programming using Function Approximators*. CRC Press, 2010.

[10] T. Dierks and S. Jagannathan. Output feedback control of a quadrotor UAV using neural networks. *IEEE Transactions on Neural Networks*, 21(1):50–66, 2010.

[11] K. Doya. Reinforcement learning in continuous time and space. *Neural Computation*, 12(1):219–245, 2000.

[12] L. A. Feldkamp and D. V. Prokhorov. Recurrent neural networks for state estimation. In: Proceedings of the Twelfth Yale Workshop on Adaptive and Learning Systems, pp. 17–22, New Haven, CT, 2003.

[13] C. M. Harris and D. M. Wolpert. Signal-dependent noise determines motor planning. *Nature*, 394:780–784, 1998.

[14] H. He, Z. Ni, and J. Fu. A three-network architecture for on-line learning and optimization based on adaptive dynamic programming. *Neurocomputing*, 78(1):3–13, 2012.

[15] R. Howard. *Dynamic Programming and Markov Processes*. MIT Press, Cambridge, MA, 1960.

[16] Y. Jiang and Z. P. Jiang. *RADPBook*. http://yu-jiang.github.io/radpbook/. Accessed May 1, 2015.

[17] Y. Jiang and Z. P. Jiang. Robust approximate dynamic programming and global stabilization with nonlinear dynamic uncertainties. In: Proceedings of the 50th IEEE Conference on Joint Decision and Control Conference and European Control Conference (CDC-ECC), pp. 115–120, Orlando, FL, 2011.

[18] Y. Jiang and Z. P. Jiang. Computational adaptive optimal control for continuous-time linear systems with completely unknown dynamics. *Automatica*, 48(10):2699–2704, 2012.

[19] Y. Jiang and Z. P. Jiang. Robust adaptive dynamic programming. In: D. Liu and F. Lewis, editors, *Reinforcement Learning and Adaptive Dynamic Programming for Feedback Control*, Chapter 13, pp. 281–302. John Wiley & Sons, 2012.

[20] Y. Jiang and Z. P. Jiang. Robust adaptive dynamic programming for large-scale systems with an application to multimachine power systems. *IEEE Transactions on Circuits and Systems II: Express Briefs*, 59(10):693–697, 2012.

[21] Y. Jiang and Z. P. Jiang. Robust adaptive dynamic programming with an application to power systems. *IEEE Transactions on Neural Networks and Learning Systems*, 24(7):1150–1156, 2013.

[22] Y. Jiang and Z. P. Jiang. Robust adaptive dynamic programming and feedback stabilization of nonlinear systems. *IEEE Transactions on Neural Networks and Learning Systems*, 25(5):882–893, 2014.

[23] Y. Jiang and Z. P. Jiang. Global adaptive dynamic programming for continuous-time nonlinear systems. *IEEE Transactions on Automatic Control*, 60(11):2917–2929, November 2015.

[24] Z. P. Jiang and Y. Jiang. Robust adaptive dynamic programming for linear and nonlinear systems: An overview. *European Journal of Control*, 19(5):417–425, 2013.

[25] Z. P. Jiang and I. Mareels. A small-gain control method for nonlinear cascaded systems with dynamic uncertainties. *IEEE Transactions on Automatic Control*, 42(3):292–308, 1997.

[26] Z. P. Jiang and L. Praly. Design of robust adaptive controllers for nonlinear systems with dynamic uncertainties. *Automatica*, 34(7):825–840, 1998.

[27] Z. P. Jiang, A. R. Teel, and L. Praly. Small-gain theorem for ISS systems and applications. *Mathematics of Control, Signals and Systems*, 7(2):95–120, 1994.

[28] Y. H. Kim and F. L. Lewis. *High-Level Feedback Control with Neural Networks*. World Scientific, 1998.

[29] B. Kiumarsi, F. L. Lewis, H. Modares, A. Karimpour, and M.-B. Naghibi-Sistani. Reinforcement Q-learning for optimal tracking control of linear discrete-time systems with unknown dynamics. *Automatica*, 50(4):1167–1175, 2014.

[30] M. Krstic, I. Kanellakopoulos, and P. V. Kokotovic. *Nonlinear and Adaptive Control Design*. John Wiley & Sons, New York, 1995.

[31] F. L. Lewis and D. Liu. *Reinforcement Learning and Approximate Dynamic Programming for Feedback Control*. John Wiley & Sons, 2012.

[32] F. L. Lewis and K. G. Vamvoudakis. Reinforcement learning for partially observable dynamic processes: Adaptive dynamic programming using measured output data. *IEEE Transactions on Systems, Man, and Cybernetics, Part B: Cybernetics*, 41(1):14–25, 2011.

[33] F. L. Lewis and D. Vrabie. Reinforcement learning and adaptive dynamic programming for feedback control. *IEEE Circuits and Systems Magazine*, 9(3):32–50, 2009.

[34] F. L. Lewis, D. Vrabie, and V. L. Syrmos. *Optimal Control*, 3rd ed. John Wiley & Sons, New York, 2012.

[35] D. Liu and D. Wang. Optimal control of unknown nonlinear discrete-time systems using iterative globalized dual heuristic programming algorithm. In: F. Lewis and L. Derong, editors, *Reinforcement Learning and Approximate Dynamic Programming for Feedback Control*, pp. 52–77. John Wiley & Sons, 2012.

[36] T. Liu, D. J. Hill, and Z. P. Jiang. Lyapunov formulation of ISS cyclic-small-gain in continuous-time dynamical networks. *Automatica*, 47(9):2088–2093, 2011.

[37] J. Mendel and R. McLaren. Reinforcement-learning control and pattern recognition systems. In: *A Prelude to Neural Networks*, pp. 287–318. Prentice Hall Press, 1994.

[38] M. Minsky. Steps toward artificial intelligence. *Proceedings of the IRE*, 49(1):8–30, 1961.

[39] H. Modares, F. L. Lewis, and M.-B. Naghibi-Sistani. Adaptive optimal control of unknown constrained-input systems using policy iteration and neural networks. *IEEE Transactions on Neural Networks and Learning Systems*, 24(10):1513–1525, 2013.

[40] J. J. Murray, C. J. Cox, G. G. Lendaris, and R. Saeks. Adaptive dynamic programming. *IEEE Transactions on Systems, Man, and Cybernetics, Part C: Applications and Reviews*, 32(2):140–153, 2002.

[41] W. B. Powell. *Approximate Dynamic Programming: Solving the Curses of Dimensionality*. John Wiley & Sons, New York, 2007.

[42] M. L. Puterman. *Markov Decision Processes: Discrete Stochastic Dynamic Programming*, Vol. 414. John Wiley & Sons, 2009.

[43] J. Si, A. G. Barto, W. B. Powell, and D. C. Wunsch (editors). *Handbook of Learning and Approximate Dynamic Programming*. John Wiley & Sons, Inc., Hoboken, NJ, 2004.

[44] D. Silver, A. Huang, C. J. Maddison, A. Guez, L. Sifre, G. Driessche, J. Schrittwieser, I. Antonoglou, V. Panneershelvam, M. Lanctot, S. Dieleman, D. Grewe, J. Nham, N. Kalchbrenner, I. Sutskever, T. Lillicrap, M. Leach, K. Kavukcuoglu, T. Graepel, and D. Hassabis. Mastering the game of Go with deep neural networks and tree search. *Nature*, 529(7587):484–489, 2016.

[45] E. D. Sontag. Input to state stability: Basic concepts and results. In: *Nonlinear and Optimal Control Theory*, pp. 163–220. Springer, 2008.

[46] R. S. Sutton. Learning to predict by the methods of temporal differences. *Machine learning*, 3(1):9–44, 1988.

[47] R. S. Sutton and A. G. Barto. *Reinforcement Learning: An Introduction*. Cambridge University Press, 1998.

[48] C. Szepesvari. Reinforcement learning algorithms for MDPs. Technical Report TR09-13, Department of Computing Science, University of Alberta, Edmonton, CA, 2009.

[49] K. G. Vamvoudakis and F. L. Lewis. Online actor–critic algorithm to solve the continuous-time infinite horizon optimal control problem. *Automatica*, 46(5):878–888, 2010.

[50] K. G. Vamvoudakis and F. L. Lewis. Multi-player non-zero-sum games: Online adaptive learning solution of coupled Hamilton–Jacobi equations. *Automatica*, 47(8):1556–1569, 2011.

[51] K. G. Vamvoudakis and F. L. Lewis. Online solution of nonlinear two-player zero-sum games using synchronous policy iteration. *International Journal of Robust and Nonlinear Control*, 22(13):1460–1483, 2012.

[52] D. Vrabie, O. Pastravanu, M. Abu-Khalaf, and F. Lewis. Adaptive optimal control for continuous-time linear systems based on policy iteration. *Automatica*, 45(2):477–484, 2009.

[53] D. Vrabie, K. G. Vamvoudakis, and F. L. Lewis. *Optimal Adaptive Control and Differential Games by Reinforcement Learning Principles*. IET, London, 2013.

[54] M. Waltz and K. Fu. A heuristic approach to reinforcement learning control systems. *IEEE Transactions on Automatic Control*, 10(4):390–398, 1965.

[55] F.-Y. Wang, H. Zhang, and D. Liu. Adaptive dynamic programming: An introduction. *IEEE Computational Intelligence Magazine*, 4(2):39–47, 2009.

[56] C. Watkins. Learning from delayed rewards. PhD Thesis, King's College of Cambridge, 1989.

[57] Q. Wei and D. Liu. Data-driven neuro-optimal temperature control of water gas shift reaction using stable iterative adaptive dynamic programming. *IEEE Transactions on Industrial Electronics*, 61(11):6399–6408, November 2014.

[58] P. Werbos. The elements of intelligence. *Cybernetica (Namur)*, (3), 1968.

[59] P. Werbos. Advanced forecasting methods for global crisis warning and models of intelligence. *General Systems Yearbook*, 22:25–38, 1977.

[60] P. Werbos. Reinforcement learning and approximate dynamic programming (RLADP) – Foundations, common misconceptions and the challenges ahead. In: F. L. Lewis and D. Liu, editors, *Reinforcement Learning and Approximate Dynamic Programming for Feedback Control*, pp. 3–30. John Wiley & Sons, Hoboken, NJ, 2013.

[61] P. J. Werbos. Beyond regression: New tools for prediction and analysis in the behavioral sciences. PhD Thesis, Harvard University, 1974.

[62] P. J. Werbos. Neural networks for control and system identification. In: Proceedings of the 28th IEEE Conference on Decision and Control, pp. 260–265, Tampa, FL, 1989.

[63] P. J. Werbos. A menu of designs for reinforcement learning over time. In: W. Miller, R. Sutton, and P. Werbos, editors, *Neural Networks for Control*, pp. 67–95. MIT Press, Cambridge, MA, 1990.

[64] P. J. Werbos. Approximate dynamic programming for real-time control and neural modeling. In: D. White and D. Sofge, editors, *Handbook of Intelligent Control: Neural, Fuzzy, and Adaptive Approaches*, pp. 493–525. Van Nostrand Reinhold, New York, 1992.

[65] P. J. Werbos. From ADP to the brain: Foundations, roadmap, challenges and research priorities. In: 2014 International Joint Conference on Neural Networks (IJCNN), pp. 107–111, Beijing, 2014. doi: 10.1109/IJCNN.2014.6889359

[66] H. Xu, S. Jagannathan, and F. L. Lewis. Stochastic optimal control of unknown linear networked control system in the presence of random delays and packet losses. *Automatica*, 48(6):1017–1030, 2012.

[67] X. Xu, C. Wang, and F. L. Lewis. Some recent advances in learning and adaptation for uncertain feedback control systems. *International Journal of Adaptive Control and Signal Processing*, 28(3–5):201–204, 2014.

[68] H. Zhang, D. Liu, Y. Luo, and D. Wang. *Adaptive Dynamic Programming for Control*. Springer, London, 2013.

[69] H. Zhang, Q. Wei, and D. Liu. An iterative adaptive dynamic programming method for solving a class of nonlinear zero-sum differential games. *Automatica*, 47(1):207–214, 2011.

[70] X. Zhang, H. He, H. Zhang, and Z. Wang. Optimal control for unknown discrete-time nonlinear Markov jump systems using adaptive dynamic programming. *IEEE Transactions on Neural Networks and Learning Systems*, 25(12):2141–2155, 2014.

CHAPTER 2

ADAPTIVE DYNAMIC PROGRAMMING FOR UNCERTAIN LINEAR SYSTEMS

This chapter presents a reinforcement learning-inspired ADP approach for finding a new class of online adaptive optimal controllers for uncertain linear systems. The only information required for feedback controller design is the dimension of the state vector $x(t)$, the dimension of the input vector $u(t)$, and an a priori linear state-feedback control policy that asymptotically stabilizes the system at the origin. The block diagram of such a setting is shown in Figure 2.1. The proposed approach employs the idea of ADP to iteratively solve the algebraic Riccati equation (ARE) using the online information of the state and the input, without requiring the a priori knowledge of, or identifying the system matrices.

2.1 PROBLEM FORMULATION AND PRELIMINARIES

Consider a continuous-time linear system described by

$$\dot{x} = Ax + Bu \tag{2.1}$$

where $x \in \mathbb{R}^n$ is the system state fully available for feedback control design; $u \in \mathbb{R}^m$ is the control input; and $A \in \mathbb{R}^{n \times n}$ and $B \in \mathbb{R}^{n \times m}$ are uncertain constant matrices. In addition, the system is assumed to be stabilizable, in the sense that there exists a constant matrix K of appropriate dimensions so that $A - BK$ is Hurwitz (i.e., all eigenvalues of $A - BK$ are located in the open-left-half plane).

Robust Adaptive Dynamic Programming, First Edition. Yu Jiang and Zhong-Ping Jiang.
© 2017 by The Institute of Electrical and Electronics Engineers, Inc. Published 2017 by John Wiley & Sons, Inc.

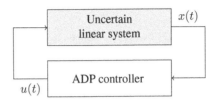

FIGURE 2.1   ADP-based online learning control for uncertain linear systems.

We are interested in finding a linear quadratic regulator (LQR) in the form of

$$u = -Kx \qquad (2.2)$$

which minimizes the following performance index

$$J(x_0; u) = \int_0^\infty (x^T Q x + u^T R u) dt \qquad (2.3)$$

where $Q = Q^T \geq 0, R = R^T > 0$, with $(A, Q^{1/2})$ observable.

According to the conventional LQR optimal control theory, with (2.2), we have

$$J(x_0; u) = x_0^T P x_0 \qquad (2.4)$$

where

$$P = \int_0^\infty e^{(A-BK)^T t} (x^T Q x + K^T R K) e^{(A-BK)t} dt \qquad (2.5)$$

is a finite matrix if and only if $A - BK$ is Hurwitz. Taking the derivative of $x^T P x$ along the solution of (2.1), it follows that P is the unique positive definite solution to the Lyapunov equation

$$(A - BK)^T P + P(A - BK) + Q + K^T R K = 0 \qquad (2.6)$$

The optimal solution to the above-mentioned problem is associated with the following well-known ARE (see [29])

$$A^T P + PA + Q - PBR^{-1}B^T P = 0 \qquad (2.7)$$

which has a unique real symmetric, positive definite solution P^*. Once P^* is obtained, the optimal feedback gain matrix K^* in (2.2) is thus determined by

$$K^* = R^{-1} B^T P^* \qquad (2.8)$$

Since (2.7) is nonlinear in P, it is usually difficult to solve it analytically, especially for large-size matrices. Nevertheless, many efficient algorithms have been developed to numerically approximate the solution to (2.7). One of such algorithms is widely known as Kleinman's algorithm [27] and is recalled below.

Theorem 2.1.1 ([27]) *Let $K_0 \in \mathbb{R}^{m \times n}$ be any stabilizing feedback gain matrix (i.e., $A - BK_0$ is Hurwitz), and repeat the following steps for $k = 0, 1, \ldots$*

(1) *Solve for the real symmetric positive definite solution P_k of the Lyapunov equation*

$$A_k^T P_k + P_k A_k + Q + K_k^T R K_k = 0 \tag{2.9}$$

where $A_k = A - BK_k$.

(2) *Update the feedback gain matrix by*

$$K_{k+1} = R^{-1} B^T P_k \tag{2.10}$$

Then, the following properties hold:

(1) *$A - BK_k$ is Hurwitz,*
(2) *$P^* \leq P_{k+1} \leq P_k$,*
(3) *$\lim_{k \to \infty} K_k = K^*$, $\lim_{k \to \infty} P_k = P^*$.*

Proof: Consider the Lyapunov equation (2.9) with $k = 0$. Since $A - BK_0$ is Hurwitz, by (2.5) we know P_0 is finite and positive definite. In addition, by (2.5) and (2.9) we have

$$P_0 - P_1 = \int_0^\infty e^{A_1^T \tau} (K_0 - K_1)^T R (K_0 - K_1) e^{A_1 \tau} d\tau \geq 0 \tag{2.11}$$

Similarly, by (2.5) and (2.7) we obtain

$$P_1 - P^* = \int_0^\infty e^{A_1^T \tau} (K_1 - K^*)^T R (K_1 - K^*) e^{A_1 \tau} d\tau \geq 0 \tag{2.12}$$

Therefore, we have $P^* \leq P_1 \leq P_0$. Since P^* is positive definite and P_0 is finite, P_1 must be finite and positive definite. This implies that $A - BK_1$ is Hurwitz. Repeating the above analysis for $k = 1, 2, \ldots$ proves Properties (1) and (2) in Theorem 2.1.1.

Finally, since $\{P_k\}$ is a monotonically decreasing sequence and lower bounded by P^*, $\lim_{k \to \infty} P_k = P_\infty$ exists. By (2.9) and (2.10), $P = P_\infty$ satisfies (2.7), which has a unique solution. Therefore, $P_\infty = P^*$. The proof is thus complete. ∎

The algorithm described in Theorem 2.1.1 is in fact a policy iteration method [17] for continuous-time linear systems. Indeed, given a stabilizing gain matrix K_k, (2.9) is known as the step of *policy evaluation*, since it evaluates the cost matrix P_k

associated with the control policy. Equation (2.10), known as *policy improvement*, finds a new feedback gain K_{k+1} based on the evaluated cost matrix P_k.

According to Theorem 2.1.1, by iteratively solving for P_k from the Lyapunov equation (2.9) and updating K_k according to (2.10), the (unique) solution to the nonlinear equation (2.7) is numerically approximated. However, in each iteration, the perfect knowledge of A and B is required, because these two matrices appear explicitly in (2.9) and (2.10). In Section 2.2, we will show how this policy iteration can be implemented via reinforcement learning, without knowing A or B, or both.

2.2 ONLINE POLICY ITERATION

To begin with, let us consider the following control policy

$$u = -K_k x + e \tag{2.13}$$

where the time-varying signal e denotes an artificial noise, known as the *exploration noise*, added for the purpose of online learning.

Remark 2.2.1   *Choosing the exploration noise is not a trivial task for general reinforcement learning problems and other related machine learning problems, especially for high-dimensional systems. In solving practical problems, several types of exploration noise have been adopted, such as random noise [1, 48], exponentially decreasing probing noise [41]. For the simulations in Section 2.4, sum of sinusoidal signals with different frequencies will be used to construct the exploration noise, as in [22].*

Under the control policy (2.13), the original system (2.1) can be rewritten as

$$\dot{x} = A_k x + Be \tag{2.14}$$

Then, taking the time derivative of $x^T P_k x$ along the solutions of (2.14), it follows that

$$\frac{d}{dt}\left(x^T P_k x\right) = x^T \left(A_k^T P_k + P_k A_k\right) x + 2e^T B^T P_k x$$

$$= -x^T Q_k x + 2e^T R K_{k+1} x \tag{2.15}$$

where $Q_k = Q + K_k^T R K_k$.

It is worth pointing out that, in (2.15) we used (2.9) to replace the term $x^T(A_k^T P_k + P_k A_k)x$, which depends on A and B by the term $-x^T Q_k x$. This new term can be measured online from real-time data along the system trajectories. Also, by (2.10), we replaced the term $B^T P_k$ with $R K_{k+1}$, in which K_{k+1} is treated as another unknown matrix to be solved together with P_k. Therefore, we have removed the dependencies on the system matrices A and B in (2.15), such that it becomes possible to solve simultaneously for P_k and K_{k+1} using online measurements.

Now, by integrating both sides of (2.15) on any given interval $[t, t + \delta t]$ and by rearranging the terms, we have

$$x^T(t + \delta t)P_k x(t + \delta t) - x^T(t)P_k x(t)$$

$$-2 \int_t^{t+\delta t} e^T R K_{k+1} x d\tau$$

$$= - \int_t^{t+\delta t} x^T Q_k x d\tau \tag{2.16}$$

We call (2.16) the *online policy iteration* equation, for it relies on the knowledge of state measurements and the control policy being applied, instead of the system knowledge. Further, we can use (2.16) to obtain a set of equations, by specifying $t = t_{k,1}, t_{k,2}, \ldots, t_{k,l_k}$, with $0 \le t_{k,i} + \delta t \le t_{k,i+1}$ and $t_{k,i} + \delta t \le t_{k+1,1}$ for all $k = 0, 1, \ldots$ and $i = 1, 2, \ldots, l_k$. These equations based on input/state data can then be used to solve for P_k and K_{k+1}. The details will be given in Section 2.3.

We also consider the online policy iteration (2.16) as an *on-policy learning* method. This is because each time when a new control policy, represented by the gain matrix K_k, is obtained, it must be implemented to generate new solutions of the closed-loop system. These new solutions are then used for evaluating the current cost and finding the new policy. To be more specific, $x(t)$ appeared in (2.16) is the solution of (2.14), in which K_k is used to formulate the control policy.

Although on-policy iteration closely mimics biological learning, the entire adaptation process can be slow and one needs to keep collecting online data until some convergence criterion is satisfied. In engineering applications, we are sometimes more interested in obtaining an approximate optimal solution by making full use of some finite data. This motivates us to develop an off-policy learning strategy, in which we apply an initial control policy to the system on a finite number of time intervals and collect the online measurements. Then, all iterations are conducted by using repeatedly the same online data.

To this end, consider the following system, which is the closed-loop system composed of (2.1) and an arbitrary feedback control policy $u = u_0$,

$$\dot{x} = Ax + Bu_0 \tag{2.17}$$

Similar to (2.16), we have

$$x^T(t + \delta t)P_k x(t + \delta t) - x^T(t)P_k x(t)$$

$$-2 \int_t^{t+\delta t} (K_k x + u_0)^T R K_{k+1} x d\tau$$

$$= - \int_t^{t+\delta t} x^T Q_k x d\tau \tag{2.18}$$

Although (2.16) and (2.18) share a very similar structure, a fundamental difference between them is that $x(t)$ in (2.18) is generated from system (2.17), in which K_k is not involved. Therefore, the same amount of data collected on the interval $[t_1, t_l + \delta t]$ can be used for calculating K_k, with $k = 1, 2, \dots$ As a result, we call this implementation *off-policy learning*, in that the actual policy been used can be an arbitrary one, as long as it keeps the solutions of the overall system bounded. The on-policy and the off-policy implementations will be further discussed in the Section 2.3.

2.3 LEARNING ALGORITHMS

2.3.1 On-Policy Learning

For computational simplicity, we would like to convert the unknown matrices P_k and K_{k+1} into a vector. One convenient way to achieve this conversion, without losing any information, is via Kronecker product representation [28]. One important identity we use is

$$\text{vec}(XYZ) = (Z^T \otimes X)\text{vec}(Y) \tag{2.19}$$

Therefore, we have

$$x^T P_k x = (x^T \otimes x^T)\text{vec}(P_k) \tag{2.20}$$

$$e^T R K_{k+1} x = (x^T \otimes (e^T R))\text{vec}(K_{k+1}) \tag{2.21}$$

Applying the above equalities (2.20)–(2.21), (2.16) can be converted to

$$\left[x^T \otimes x^T |_t^{t+\delta t} \quad -2 \int_t^{t+\delta t}(x^T \otimes (e^T R))dt \right] \begin{bmatrix} \text{vec}(P_k) \\ \text{vec}(K_{k+1}) \end{bmatrix}$$

$$= -\int_t^{t+\delta t} x^T Q_k x \, dt \tag{2.22}$$

As mentioned in Section 2.2, we now can apply (2.16) on multiple time intervals to obtain a set of linear equations represented in the following matrix form

$$\Theta_k \begin{bmatrix} \text{vec}(P_k) \\ \text{vec}(K_{k+1}) \end{bmatrix} = \Xi_k \tag{2.23}$$

where

$$\Theta_k = \begin{bmatrix} x^T \otimes x^T |_{t_{k,1}}^{t_{k,1}+\delta t} & -2\int_{t_{k,1}}^{t_{k,1}+\delta t}(x^T \otimes e^T R)dt \\ x^T \otimes x^T |_{t_{k,2}}^{t_{k,2}+\delta t} & -2\int_{t_{k,2}}^{t_{k,2}+\delta t}(x^T \otimes e^T R)dt \\ \vdots & \vdots \\ x^T \otimes x^T |_{t_{k,l_k}}^{t_{k,l_k}+\delta t} & -2\int_{t_{k,l_k}}^{t_{k,l_k}+\delta t}(x^T \otimes e^T R)dt \end{bmatrix}, \tag{2.24}$$

$$\Xi_k = \begin{bmatrix} -\int_{t_{k,1}}^{t_{k,1}+\delta t} x^T Q_k x dt \\ -\int_{t_{k,2}}^{t_{k,2}+\delta t} x^T Q_k x dt \\ \vdots \\ -\int_{t_{k,l_k}}^{t_{k,l_k}+\delta t} x^T Q_k x dt \end{bmatrix} \tag{2.25}$$

Before solving the pair (P_k, K_{k+1}) from (2.23), it is important to check if the solution is unique. To this end, Assumption 2.3.1 is introduced.

Assumption 2.3.1 *For each $k = 0, 1, 2, \ldots$, there exists a sufficiently large integer $l_k > 0$, such that the following rank condition holds.*

$$\text{rank}(\Theta_k) = \frac{n(n+1)}{2} + mn \tag{2.26}$$

Each interval $[t_{k,j}, t_{k,j+1}]$ is called a sampling interval. We need to collect enough sampled data (which means large enough l_k for each iteration step k). The choice of the exploration noise plays a vital role. In general, this rank condition can be checked computationally, but not analytically.

To satisfy the rank condition in (2.26), a good practice is to assure that one iteration step can utilize data from at least twice as many sampling intervals as the unknowns, that is, $l_k \geq n(n+1) + 2mn$ for $k = 0, 1, 2, \cdots$. In addition, if the exploration noise is a periodical signal, the length of any sampling interval should be sufficiently larger than the period of the noise.

Remark 2.3.2 *The rank condition (2.26) introduced above is essentially inspired from the persistent excitation (PE) condition in adaptive control [18, 33].*

Lemma 2.3.3 *Under Assumption 2.3.1, there is a unique pair $(P_k, K_{k+1}) \in \mathbb{R}^{n \times n} \times \mathbb{R}^{n \times m}$ satisfying (2.23) with $P_k = P_k^T$.*

Proof: By (2.24), there are $\frac{n(n-1)}{2}$ duplicated columns in the first n^2 columns of Θ_k. Since P_k is symmetric, there are $\frac{n(n-1)}{2}$ duplicated entries in the first n^2 entries of the vector $\begin{bmatrix} \text{vec}(P_k) \\ \text{vec}(K_{k+1}) \end{bmatrix}$, and the row indices of these duplicated entries match exactly the indices of the $\frac{n(n-1)}{2}$ duplicated columns in Θ_k. For example, if $n = 2$, the third column of Θ_k is the duplicated column because it is identical to the second column of Θ_k. Meanwhile, the third entry in $\begin{bmatrix} \text{vec}(P_k) \\ \text{vec}(K_{k+1}) \end{bmatrix}$ is duplicated from the second entry in the same vector.

Under Assumption 2.3.1, the $\frac{n(n+1)}{2} + mn$ distinct columns in Θ_k are linearly independent. As discussed above, the indices of these independent columns are exactly the same as the row indices of the $\frac{n(n+1)}{2} + mn$ distinct elements in the vector

$\begin{bmatrix} \text{vec}(P_k) \\ \text{vec}(K_{k+1}) \end{bmatrix}$, provided that P_k is symmetric. Therefore, the pair (P_k, K_{k+1}) satisfying (2.23) with $P_k = P_k^T$ must be unique. ∎

As long as the rank condition (2.26) is satisfied, the unique pair (P_k, K_{k+1}) mentioned in Lemma 2.3.3 can be easily solved in MATLAB. In particular, the following MATLAB function can be used. The four input arguments of this function are expected to be Θ_K, Ξ_k, m, and n, respectively. The two output arguments are the corresponding matrices P_k and K_{k+1}.

```
function [P,K] = PKsolver(Theta,Xi,m,n)
  w = pinv(Theta)*Xi;             % Solve for w = [vec(P); vec(K)]
  P = reshape(w(1:n*n),n,n);      % Reshape w(1:n^2) to get P
  P = (P + P')/2;                 % Convert P to a symmetric matrix
  K = reshape(w(n*n+1:end),m,n);  % Reshape w(n^2+1:end) to get K
end
```

Now, we are ready to give the following on-policy continuous-time ADP algorithm, a flowchart describing the algorithm is shown in Figure 2.2.

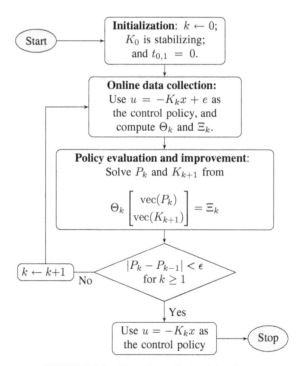

FIGURE 2.2   Flowchart of Algorithm 2.3.4.

Algorithm 2.3.4   *On-policy ADP algorithm*

(1) *Initialization:*
    Find K_0 such that $A - BK_0$ is Hurwitz. Let $k = 0$ and $t_{0,1} = 0$.
(2) *Online data collection:*
    Apply $u = -K_k x + e$ to the system from $t = t_{k,1}$ and construct each row of the the data matrices Θ_k and Ξ_k, until the rank condition (2.26) is satisfied.
(3) *Policy evaluation and improvement:*
    Solve for $P_k = P_k^T$ and K_{k+1} from (2.23).
(4) *Stopping criterion:*
    Terminate the exploration noise and apply $u = -K_k x$ as the control, if $k \geq 1$ and

$$|P_k - P_{k-1}| < \epsilon \tag{2.27}$$

with $\epsilon > 0$ a predefined, sufficiently small, threshold. Otherwise, let $t_{k+1,1}$ satisfy $t_{k+1,1} \geq t_{k,l_k} + \delta t$ and go to Step (2), with $k \leftarrow k + 1$.

Remark 2.3.5   *Notice that $\epsilon > 0$ is selected to balance the exploration/exploitation trade-off. In practice, a larger ϵ may lead to shorter exploration time and therefore will allow the system to implement the control policy and terminate the exploration noise sooner. On the other hand, to obtain more accurate approximation of the optimal solution, the threshold ϵ is chosen small, and (2.27) should hold for several consecutive values of k. The same is true for off-policy algorithms.*

Theorem 2.3.6   *Let $K_0 \in \mathbb{R}^{m \times n}$ be any stabilizing feedback gain matrix, and let (P_k, K_{k+1}) be a pair of matrices obtained from Algorithm 2.3.4. Then, under Assumption 2.3.1, the following properties hold:*

(1) *$A - BK_k$ is Hurwitz,*
(2) *$P^* \leq P_{k+1} \leq P_k$,*
(3) *$\lim_{k \to \infty} K_k = K^*$, $\lim_{k \to \infty} P_k = P^*$.*

Proof: From (2.15), (2.16), and (2.22), one sees that the pair (P_k, K_{k+1}) obtained from (2.9) and (2.10) must satisfy (2.23). In addition, by Lemma 2.3.3, such a pair is unique. Therefore, the solution to (2.9) and (2.10) is the same as the solution to (2.23) for all $k = 0, 1, \ldots$. The proof is thus completed by Theorem 2.1.1.   ■

Remark 2.3.7   *The ADP approach introduced here is related to the action-dependent heuristic dynamic programming (ADHDP) [46], or Q-learning [44] method for discrete-time systems. Indeed, it can be viewed that we solve for the following matrix H_k at each iteration step.*

$$H_k = \begin{bmatrix} H_{11,k} & H_{12,k} \\ H_{21,k} & H_{22,k} \end{bmatrix} = \begin{bmatrix} P_k & P_k B \\ B^T P_k & R \end{bmatrix} \tag{2.28}$$

Once this matrix is obtained, the control policy can be updated by

$$K_{k+1} = H_{22,k}^{-1} H_{21,k} \tag{2.29}$$

The discrete-time version of the H_k matrix can be found in [5] and [31].

2.3.2 Off-Policy Learning

As mentioned earlier, the off-policy learning approach aims to find an approximate optimal control policy by using the online measurements of the closed-loop system (2.17). Notice that, if we directly compute the data matrices Θ_k and Ξ_k as in the on-policy learning, we cannot reuse these two data matrices for subsequent iterations, because these two matrices both depend on K_k. Therefore, it is desirable that we can find a way to compute Θ_k and Ξ_k for all $k = 0, 1, \ldots$, by repeatedly using certain online measurements.

To this end, we apply (2.20) again to obtain the following equalities:

$$x^T Q_k x = (x^T \otimes x^T) \text{vec}(Q_k) \tag{2.30}$$

and

$$
(u + K_k x)^T R K_{k+1} x
$$
$$
= \left[(x^T \otimes x^T)(I_n \otimes K_k^T R) + (x^T \otimes u_0^T)(I_n \otimes R) \right] K_{k+1} \tag{2.31}
$$

Further, for any positive integer l, we define matrices $\delta_{xx} \in \mathbb{R}^{l \times n^2}$, $I_{xx} \in \mathbb{R}^{l \times n^2}$, and $I_{xu} \in \mathbb{R}^{l \times mn}$, such that

$$
\delta_{xx} = \left[x \otimes x|_{t_1}^{t_1+\delta t}, \quad x \otimes x|_{t_2}^{t_2+\delta t}, \quad \ldots, \quad x \otimes x|_{t_l}^{t_l+\delta t} \right]^T \tag{2.32}
$$

$$
I_{xx} = \left[\int_{t_1}^{t_1+\delta t} x \otimes x d\tau, \quad \int_{t_2}^{t_2+\delta t} x \otimes x d\tau, \quad \ldots, \quad \int_{t_l}^{t_l+\delta t} x \otimes x d\tau \right]^T \tag{2.33}
$$

$$
I_{xu} = \left[\int_{t_1}^{t_1+\delta t} x \otimes u_0 d\tau, \quad \int_{t_2}^{t_2+\delta t} x \otimes u_0 d\tau, \quad \ldots, \quad \int_{t_l}^{t_l+\delta t} x \otimes u_0 d\tau \right]^T \tag{2.34}
$$

where $0 \le t_1 < t_2 < \cdots < t_l$.

Then, for any given stabilizing gain matrix K_k, (2.18) implies the following matrix form of linear equations

$$
\tilde{\Theta}_k \begin{bmatrix} \text{vec}(P_k) \\ \text{vec}(K_{k+1}) \end{bmatrix} = \tilde{\Xi}_k \tag{2.35}
$$

where $\tilde{\Theta}_k \in \mathbb{R}^{l \times (n^2 + mn)}$ and $\tilde{\Xi}_k \in \mathbb{R}^l$ are defined as:

$$\tilde{\Theta}_k = \left[\delta_{xx}, -2I_{xx}(I_n \otimes K_k^T R) - 2I_{xu}(I_n \otimes R)\right] \tag{2.36}$$

$$\tilde{\Xi}_k = -I_{xx}\text{vec}(Q_k) \tag{2.37}$$

As can be seen from (2.32)–(2.34), (2.36), and (2.37), an initial stabilizing control policy u_0 is applied and the online information is recorded in matrices δ_{xx}, I_{xx}, and I_{xu}. Then, without requiring additional system information, the matrices δ_{xx}, I_{xx}, and I_{xu} can be repeatedly used to construct the data matrices $\tilde{\Theta}_k$ and $\tilde{\Xi}_k$, which will later be used for iterations.

Similar to the on-policy learning, we introduce the following Assumption that helps to assure the uniqueness of the solution to (2.35).

Assumption 2.3.8 *There exists a sufficiently large integer $l > 0$, such that*

$$\text{rank}([\,I_{xx} \quad I_{xu}\,]) = \frac{n(n+1)}{2} + mn \tag{2.38}$$

Lemma 2.3.9 *Under Assumption 2.3.8, there is a unique pair of matrices (P_k, K_{k+1}), with $P_k = P_k^T$, such that*

$$\tilde{\Theta}_k \begin{bmatrix} \text{vec}(P_k) \\ \text{vec}(K_{k+1}) \end{bmatrix} = \tilde{\Xi}_k \tag{2.39}$$

Proof: We only need to show that, given $Y = Y^T \in \mathbb{R}^{n \times n}$ and $Z \in \mathbb{R}^{m \times n}$, the following matrix equality holds only if $Y = 0$ and $Z = 0$

$$\tilde{\Theta}_k \begin{bmatrix} \text{vec}(Y) \\ \text{vec}(Z) \end{bmatrix} = 0 \tag{2.40}$$

It is easy to see

$$\tilde{\Theta}_k \begin{bmatrix} \text{vec}(Y) \\ \text{vec}(Z) \end{bmatrix} = [\,I_{xx} \quad 2I_{xu}\,] \begin{bmatrix} \text{vec}(M) \\ \text{vec}(N) \end{bmatrix} \tag{2.41}$$

where

$$M = A_k^T Y + YA_k + K_k^T(B^T Y - RZ) + (YB - Z^T R)K_k \tag{2.42}$$

$$N = B^T Y - RZ \tag{2.43}$$

Notice that since M is symmetric, there are in total $\frac{n(n+1)}{2} + mn$ distinct entries in M and N. The row indices of these distinct entries in the vector $\begin{bmatrix} \text{vec}(Y) \\ \text{vec}(Z) \end{bmatrix}$ match

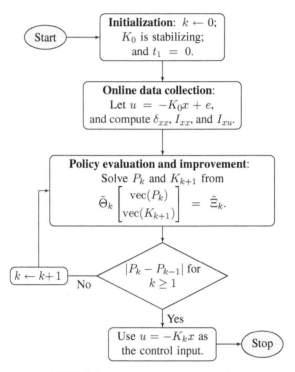

FIGURE 2.3   Flowchart of Algorithm 2.3.10.

exactly the column indices of the $\frac{n(n+1)}{2} + mn$ distinct columns in $[\, I_{xx} \quad I_{xu} \,]$. Under the rank condition (2.38) in Assumption 2.3.8, and by (2.40) and (2.41), we have $M = 0$ and $N = 0$.

Now, by (2.43) we know $Z = R^{-1}B^T Y$, and (2.42) is reduced to the following Lyapunov equation

$$A_k^T Y + Y A_k = 0 \tag{2.44}$$

Since A_k is Hurwitz for all $k \in \mathbb{Z}_+$, the only solution to (2.44) is $Y = 0$. Finally, by (2.43) we have $Z = 0$.

The proof is complete. ∎

The off-policy ADP algorithm is given below, and a flowchart is provided in Figure 2.3.

Algorithm 2.3.10   *Off-policy ADP algorithm*

(1) *Initialization:*
    Find K_0 such that $A - BK_0$ is Hurwitz. Let $k = 0$.

(2) *Online data collection:*
   Apply $u_0 = -K_0 x + e$ as the control input from $t = t_1 = 0$, where K_0 is stabilizing and e is the exploration noise. Compute δ_{xx}, I_{xx}, and I_{xu} until the rank condition in (2.38) is satisfied. Let $k = 0$.

(3) *Policy evaluation and improvement:*
   Solve P_k, with $P_k = P_k^T$, and K_{k+1} from (2.35).

(4) *Off-policy iteration:*
   Let $k \leftarrow k + 1$, and repeat Step (3), until $|P_k - P_{k-1}| \leq \epsilon$ for $k \geq 1$, where the constant $\epsilon > 0$ is a predefined small threshold.

(5) *Exploitation:*
   Use $u = -K_k x$ as the approximate optimal control policy.

Remark 2.3.11   *The initial control policy u_0 can take different forms other than $u_0 = -K_0 x + e$. For example, it can be a nonlinear feedback control policy, or even an open-loop control policy. However, it must guarantee that the system trajectories are bounded and in particular, no finite-escape occurs before switching to the approximated optimal control policy. In addition, it must be selected such that the rank condition described in Assumption 2.3.8 is satisfied.*

The convergence property of the off-policy learning algorithm 2.3.10 is summarized in the following theorem, of which the proof is omitted because it is nearly identical to the proof of Theorem 2.3.6 .

Theorem 2.3.12   *Let $K_0 \in \mathbb{R}^{m \times n}$ be any stabilizing feedback gain matrix, and let (P_k, K_{k+1}) be a pair of matrices obtained from Algorithm 2.3.10. Then, under Assumption 2.3.8, the following properties hold:*

(1) $A - BK_k$ is Hurwitz,
(2) $P^* \leq P_{k+1} \leq P_k$,
(3) $\lim\limits_{k \to \infty} K_k = K^*$, $\lim\limits_{k \to \infty} P_k = P^*$.

2.3.3 On-Policy Learning versus Off-Policy Learning

Comparing the on-policy and the off-policy learning strategies, it can be seen that the former spreads the computational burden into different iteration time points, at the price of a longer learning process. The latter can achieve faster learning by making full use of the online measurements, at the expense of heavier computational efforts at a single iteration time point. In addition, if any of the plant parameters suddenly changed to a different value, the on-policy learning method can gradually adapt to the new environment, while the off-policy learning cannot adapt automatically because it uses the data collected on some finite time intervals from the original dynamic processes. In other words, each time the system parameters change, we may need to manually restart the off-policy learning process.

Fortunately, these two learning methods are not mutually exclusive, but can be effectively integrated. For example, if the plant parameters are slowly but piecewise varying, one can periodically turn on the off-policy learning to train the controller of the system. Each time when the off-policy learning algorithm is re-applied, the initial control policy can be specified as the sum of the control policy obtained from the previous off-policy learning and exploration noise.

It is worth noticing that, under certain circumstances, on-policy learning and off-policy learning are not both applicable, as will be shown in the remainder of this book. For example, off-policy design is much easier for neural network-based ADP, as will be seen in Chapter 3. This is because implementing a new nonlinear control policy at each iteration step will result in a different area of attraction of the closed-loop system, and will therefore bring difficulties for convergence and stability analysis. In Chapter 7, we will solely use on-policy learning because it models better the adaptation mechanism of the central nervous system of humans.

2.4 APPLICATIONS

In this section, we use two examples to validate, through numerical simulations, the effectiveness of the proposed algorithms. The first example is created using Simulink Version 8.5 (R2015a). It illustrates how the on-policy learning can be applied to a third-order linear system. The second example is implemented in MATLAB scripts and the simulation was made in MATLAB R2015a. It applies the off-policy learning strategy to design an approximate optimal feedback control policy for a diesel engine [25], using online data measured from the open-loop system. All of the source files that can be used to reproduce the simulations are available in [19].

2.4.1 Example 1: On-Policy Learning for a Third-Order System

Consider the following linear system

$$\dot{x} = \begin{bmatrix} 0 & 1 & 0 \\ 0 & 0 & 1 \\ -a_1 & -a_2 & -a_3 \end{bmatrix} x + \begin{bmatrix} 0 \\ 0 \\ 1 \end{bmatrix} u \tag{2.45}$$

where a_1, a_2, and a_3 are uncertain constants but the uncontrolled system is assumed to be always asymptotically stable, that is, the three poles of the system are located in the left-half-plane.

The design objective is to find a controller that minimizes the following cost functional

$$J(x_0; u) = \int_0^\infty (|x|^2 + u^2)dt \tag{2.46}$$

To begin with, all the relevant parameters used in the algorithm are configured as follows: the initial stabilizing feedback gain is set to be $K = [0 \quad 0 \quad 0]$; the length of each learning interval is 0.1, i.e., $\delta t = 0.1$ and $t_{k+1,1} - t_{k,l_k} = t_{k,i+1} - t_{k,i} = 0.1$; the stopping threshold is set to be $\epsilon = 0.5 \times 10^{-4}$; we use $\texttt{ode45}$ as the *variable-step solver*, with the maximum step size set as 0.001; the exploration noise is selected to be a sinusoidal signal $e = 2\sin(100t)$. As a result, each iteration relies on the data collected from 10 sampling intervals, the length of which is significantly larger than the period of the exploration noise. Only for simulation purpose, we set $a_1 = 0.1$, $a_2 = 0.5$, and $a_3 = 0.7$.

The top level structure of the Simulink model is shown in Figure 2.4. The model is simulated from $t = 0$ to $t = 10$, and convergence is attained around $t = 6.5$ after 6 iterations are made. The system trajectories are plotted in Figure 2.5. The improvement of the feedback gains is shown in Figure 2.6. The cost matrix we obtained is shown in the upper-left *display* block. For comparison purpose, the optimal cost matrix is given below.

$$P^* = \begin{bmatrix} 2.3550 & 2.2385 & 0.9050 \\ 2.2385 & 4.2419 & 1.8931 \\ 0.9050 & 1.8931 & 1.5970 \end{bmatrix} \tag{2.47}$$

In addition, the optimal gain matrix is given below.

$$K^* = [\,0.9050 \quad 1.8931 \quad 1.5970\,] \tag{2.48}$$

One can compare these values with the simulation results shown in Figure 2.4.

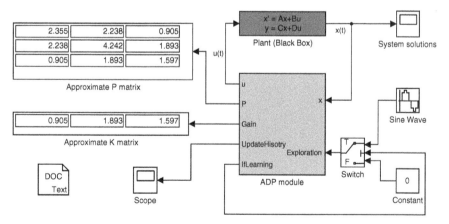

Adaptive Dynamic Programming for a third-order Continuous-Time Linear System

Copyright 2012-2015 Yu Jiang and Zhong-Ping Jiang

FIGURE 2.4   Layout of the Simulink model. The uncertain plant is implemented using a State-Space block, while the ADP algorithm is implemented in the masked subsystem *ADP module*, which contains only basic Simulink blocks. In the ADP module, an initial feedback gain matrix, the weighting matrices Q and R, the length of the learning interval, and the convergence threshold ϵ are specified as the mask parameters. The model is available at [19].

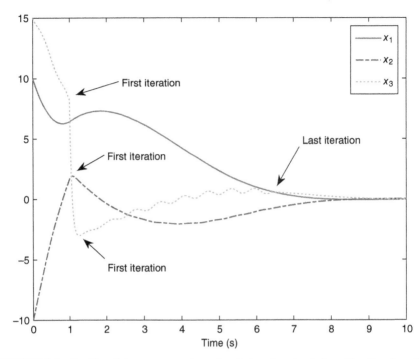

FIGURE 2.5 Profile of the system trajectories during the simulation. The first iteration occurred around $t = 1$s and the final iteration happened at $t = 6.5$s.

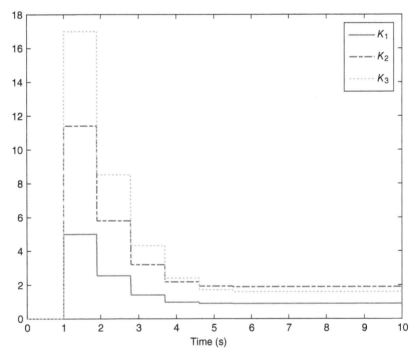

FIGURE 2.6 Profile of the feedback gains during the simulation.

2.4.2 Example 2: Off-Policy Learning for a Turbocharged Diesel Engine

Consider a diesel engine with exhaust gas re-circulation, modeled by a sixth-order linear system with the following pair of (A, B) matrices [25]:

$$A = \begin{bmatrix} -0.4125 & -0.0248 & 0.0741 & 0.0089 & 0 & 0 \\ 101.5873 & -7.2651 & 2.7608 & 2.8068 & 0 & 0 \\ 0.0704 & 0.0085 & -0.0741 & -0.0089 & 0 & 0.0200 \\ 0.0878 & 0.2672 & 0 & -0.3674 & 0.0044 & 0.3962 \\ -1.8414 & 0.0990 & 0 & 0 & -0.0343 & -0.0330 \\ 0 & 0 & 0 & -359.0000 & 187.5364 & -87.0316 \end{bmatrix}$$

$$B = \begin{bmatrix} -0.0042 & 0.0064 \\ -1.0360 & 1.5849 \\ 0.0042 & 0 \\ 0.1261 & 0 \\ 0 & -0.0168 \\ 0 & 0 \end{bmatrix}$$

Again, the precise knowledge of A and B is not used in the design of optimal controllers, they are only used for simulating the diesel engine. Since we know the physical system is asymptotically stable, the initial stabilizing feedback gain K_0 can be set as a two-by-six zero matrix.

The weighting matrices are selected to be

$$Q = \begin{bmatrix} 100 & 0 & 0 & 0 & 0 & 0 \\ 0 & 0 & 0 & 0 & 0 & 0 \\ 0 & 0 & 0 & 0 & 0 & 0 \\ 0 & 0 & 0 & 0 & 0 & 0 \\ 0 & 0 & 0 & 0 & 0 & 0 \\ 0 & 0 & 0 & 0 & 0 & 100 \end{bmatrix} \quad \text{and} \quad R = \begin{bmatrix} 1 & 0 \\ 0 & 1 \end{bmatrix}$$

In the simulation, the initial values for the state variables are randomly selected around the origin. From $t = 0$s to $t = 1$s, we inject the noise $e = \sum_{k=1}^{100} \sin(\omega_k t)$, with ω_k, with $k = 1, \dots, 100$ random numbers uniformly distributed on $[-50, 50]$.

State and input information is collected over each interval of 0.1s. The policy iteration started at $t = 1$s, and convergence is attained after 16 iterations, when the stopping criterion $|P_k - P_{k-1}| \leq 0.0005$ is satisfied. The learned controller is used as the actual control input to the system starting from $t = 1$s to the end of the simulation. The convergence of $\{P_k\}$ and $\{K_k\}$ to their optimal values is illustrated in Figure 2.7. The system output $y_1 = 3.6x_6$ denoting the mass air flow (MAF) is plotted in Figure 2.8.

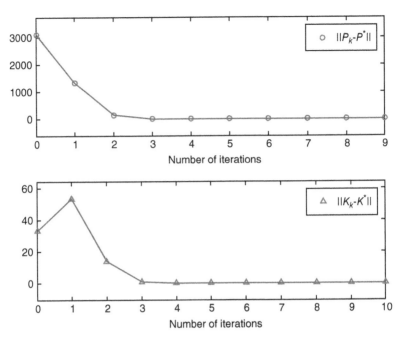

FIGURE 2.7 Convergence of P_k and K_k to their optimal values P^* and K^* during the learning process.

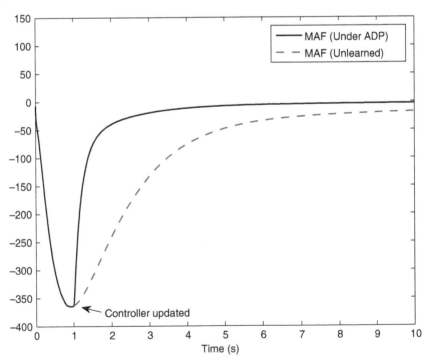

FIGURE 2.8 Profile of the MAF output during the simulation. The controller is updated at $t = 1s$.

The proposed algorithm gives the cost and the feedback gain matrices as shown below:

$$P_{10} = \begin{bmatrix} 766.7597 & 1.4018 & 97.3214 & 119.0927 & -111.2661 & 0.6265 \\ 1.4018 & 0.1032 & -0.6571 & 3.9895 & -1.8201 & 0.0150 \\ 97.3214 & -0.6571 & 71.7237 & -1.3720 & -29.0113 & 0.0230 \\ 119.0927 & 3.9895 & -1.3720 & 233.8110 & -126.0850 & -1.1828 \\ -111.2661 & -1.8201 & -29.0113 & -126.0850 & 88.1925 & 0.5859 \\ 0.6265 & 0.0150 & 0.0230 & -1.1828 & 0.5859 & 0.5687 \end{bmatrix}$$

$$K_{10} = \begin{bmatrix} 10.7537 & 0.3875 & 0.4003 & 24.8445 & -13.6682 & -0.1672 \\ 8.9982 & 0.2031 & 0.0687 & 9.2033 & -5.0784 & 0.0180 \end{bmatrix}$$

For comparison purpose, we give the optimal solutions by solving directly the ARE (2.7).

$$P^* = \begin{bmatrix} 766.7586 & 1.4020 & 97.3265 & 119.0937 & -111.2549 & 0.6265 \\ 1.4020 & 0.1032 & -0.6581 & 3.9893 & -1.8221 & 0.0150 \\ 97.3265 & -0.6581 & 71.6915 & -1.3767 & -29.0678 & 0.0230 \\ 119.0937 & 3.9893 & -1.3767 & 233.8101 & -126.0952 & -1.1828 \\ -111.2549 & -1.8221 & -29.0678 & -126.0952 & 88.0775 & 0.5859 \\ 0.6265 & 0.0150 & 0.0230 & -1.1828 & 0.5859 & 0.5687 \end{bmatrix}$$

$$K^* = \begin{bmatrix} 10.7537 & 0.3875 & 0.4006 & 24.8446 & -13.6677 & -0.1672 \\ 8.9983 & 0.2031 & 0.0681 & 9.2032 & -5.0796 & 0.0180 \end{bmatrix}$$

Notice that the problem can also be solved using the on-policy learning method described in Algorithm 2.3.4. In addition, if B is accurately known, the problem can also be solved using the method in [42]. However, on-policy learning methods would require a total learning time of 10s for 10 iterations, if the state and input information within 1 second is collected for each iteration.

2.5 NOTES

The adaptive controller design for uncertain linear systems has been intensively studied in the past literature [18, 33, 39]. A conventional way to design an indirect adaptive optimal control law can be pursued by identifying the system parameters first and then solving the related ARE. However, adaptive systems designed this way are known to respond slowly to parameter variations from the plant. There are some known direct adaptive optimal control algorithms for linear systems, which are model-based [15, 18]. It is important to notice that they are not extensible to nonlinear plants, let alone the robustness analysis to dynamic uncertainties. As will be shown in later chapters, the proposed ADP algorithm based on reinforcement learning techniques can be generalized to a wider class of genuinely nonlinear systems.

As a reinforcement learning-inspired approach, ADP theories have been studied for solving optimal control problems for uncertain systems in recent years (see, e.g., [3, 6, 8, 21, 30, 31, 42, 43, 47–49]). Among all the different ADP approaches, for discrete-time (DT) systems, the ADHDP [46], or Q-learning [44], is an online iterative scheme that does not depend on the model to be controlled, and it has found applications in many engineering disciplines [1, 30, 45, 48].

Nevertheless, due to the different structures of the AREs between DT and continuous-time (CT) systems, results developed for the DT setting cannot be directly applied for solving CT problems. In the CT setting, discretization-based methods were proposed in [2] and [7]. Alternative algorithms were developed in [36], where the state derivatives must be used. One of the first exact methods (i.e., without discretization) for CT linear systems can be found in [42], where the state derivatives and the system matrix A are not assumed to be known, but the precise knowledge of the input matrix B is required. This assumption on B has been further relaxed in [22], such that the a priori knowledge of the system dynamics can be completely unknown. The online implementation method in [22] is also considered as an off-policy method, as opposed to the on-policy methods in the past literature. The off-policy learning developed in this chapter only requires an arbitrary stabilizing control policy for the entire learning phase and can efficiently utilize the same amount of online measurements for multiple iteration steps, at the expense of a higher computational burden at a single iteration time point.

In this chapter, we have introduced the main methodology proposed in [22]. In addition, an on-policy version of the same methodology is developed. A comparison is made between these two online implementation methods, and the possible combination of them is addressed.

Finally, it is worth pointing out that the methodology developed in this chapter serves as an important computational tool to study the adaptive optimal control of CT systems. It can also be applied to many relevant linear control problems, such as output feedback adaptive optimal control [10, 20, 29], disturbance attenuation [32, 37, 38], optimal tracking [26, 34], asymptotic tracking with disturbance rejection [11, 12, 13], the optimal control of Markov jump systems [16], and games [32, 35, 40, 49]. Practical applications in power systems [4, 23, 24], helicopters, and connected vehicles are recently considered as well [9, 14]. Extensions to CT nonlinear systems and applications to engineering and computational neuroscience problems will be discussed in the next chapters.

REFERENCES

[1] A. Al-Tamimi, F. L. Lewis, and M. Abu-Khalaf. Model-free Q-learning designs for linear discrete-time zero-sum games with application to H_∞ control. *Automatica*, 43(3):473–481, 2007.

[2] L. C. Baird. Reinforcement learning in continuous time: Advantage updating. In: Proceedings of the IEEE World Congress on Computational Intelligence, Vol. 4, pp. 2448–2453, Orlando, FL, 1994.

[3] S. Bhasin, N. Sharma, P. Patre, and W. Dixon. Asymptotic tracking by a reinforcement learning-based adaptive critic controller. *Journal of Control Theory and Applications*, 9(3):400–409, 2011.

[4] T. Bian, Y. Jiang, and Z. P. Jiang. Decentralized adaptive optimal control of large-scale systems with application to power systems. *IEEE Transactions on Industrial Electronics*, 62(4):2439–2447, 2015.

[5] S. J. Bradtke, B. E. Ydstie, and A. G. Barto. Adaptive linear quadratic control using policy iteration. In: Proceedings of the American Control Conference, Vol. 3, pp. 3475–3479, Baltimore, MD, 1994.

[6] T. Dierks and S. Jagannathan. Output feedback control of a quadrotor UAV using neural networks. *IEEE Transactions on Neural Networks*, 21(1):50–66, 2010.

[7] K. Doya. Reinforcement learning in continuous time and space. *Neural Computation*, 12(1):219–245, 2000.

[8] S. Ferrari, J. E. Steck, and R. Chandramohan. Adaptive feedback control by constrained approximate dynamic programming. *IEEE Transactions on Systems, Man, and Cybernetics, Part B: Cybernetics*, 38(4):982–987, 2008.

[9] W. Gao, M. Huang, Z. P. Jiang, and T. Chai. Sampled-data-based adaptive optimal output-feedback control of a 2-degree-of-freedom helicopter. *IET Control Theory and Applications*, 10(12):1440–1447, 2016.

[10] W. Gao, Y. Jiang, Z. P. Jiang, and T. Chai. Output-feedback adaptive optimal control of interconnected systems based on robust adaptive dynamic programming. *Automatica*, 72(10):37–45, 2016.

[11] W. Gao and Z. P. Jiang. Global optimal output regulation of partially linear systems via robust adaptive dynamic programming. In: Proceedings of the 1st IFAC Conference on Modeling, Identification, and Control of Nonlinear Systems, Vol. 48, pp. 742–747, Saint Petersburg, Russia, June 2015.

[12] W. Gao and Z. P. Jiang. Linear optimal tracking control: An adaptive dynamic programming approach. In: Proceedings of the American Control Conference, pp. 4929–4934, Chicago, IL, July 2015.

[13] W. Gao and Z. P. Jiang. Adaptive dynamic programming and adaptive optimal output regulation of linear systems. *IEEE Transactions on Automatic Control*, 61(12):4164–4169, 2016.

[14] W. Gao, Z. P. Jiang, and K. Ozbay. Adaptive optimal control of connected vehicles. In: Proceedings of the 10th International Workshop on Robot Motion and Control (RoMoCo'15), pp. 288–293, Poznan, Poland, July 2015.

[15] L. Guo. On adaptive stabilization of time-varying stochastic systems. *SIAM Journal on Control and Optimization*, 28(6):1432–1451, 1990.

[16] S. He, J. Song, Z. Ding, and F. Liu. Online adaptive optimal control for continuous-time Markov jump linear systems using a novel policy iteration algorithm. *IET Control Theory & Applications*, 9(10):1536–1543, June 2015.

[17] R. Howard. *Dynamic Programming and Markov Processes*. MIT Press, Cambridge, MA, 1960.

[18] P. A. Ioannou and J. Sun. *Robust Adaptive Control.* Prentice-Hall, Upper Saddle River, NJ, 1996.

[19] Y. Jiang and Z. P. Jiang. RADP Book. http://yu-jiang.github.io/radpbook/. Accessed May 1, 2015.

[20] Y. Jiang and Z. P. Jiang. Approximate dynamic programming for output feedback control. In: Proceedings of the 29th Chinese Control Conference, pp. 5815–5820, Beijing, July 2010.

[21] Y. Jiang and Z. P. Jiang. Approximate dynamic programming for optimal stationary control with control-dependent noise. *IEEE Transactions on Neural Networks*, 22(12):2392–2398, 2011.

[22] Y. Jiang and Z. P. Jiang. Computational adaptive optimal control for continuous-time linear systems with completely unknown dynamics. *Automatica*, 48(10):2699–2704, 2012.

[23] Y. Jiang and Z. P. Jiang. Robust adaptive dynamic programming for large-scale systems with an application to multimachine power systems. *IEEE Transactions on Circuits and Systems II: Express Briefs*, 59(10):693–697, 2012.

[24] Y. Jiang and Z. P. Jiang. Robust adaptive dynamic programming with an application to power systems. *IEEE Transactions on Neural Networks and Learning Systems*, 24(7):1150–1156, 2013.

[25] M. Jung, K. Glover, and U. Christen. Comparison of uncertainty parameterisations for H_∞ robust control of turbocharged diesel engines. *Control Engineering Practice*, 13(1):15–25, 2005.

[26] B. Kiumarsi, F. L. Lewis, H. Modares, A. Karimpour, and M.-B. Naghibi-Sistani. Reinforcement Q-learning for optimal tracking control of linear discrete-time systems with unknown dynamics. *Automatica*, 50(4):1167–1175, 2014.

[27] D. Kleinman. On an iterative technique for Riccati equation computations. *IEEE Transactions on Automatic Control*, 13(1):114–115, 1968.

[28] A. J. Laub. *Matrix Analysis for Scientists and Engineers.* SIAM, 2005.

[29] F. Lewis, D. Vrabie, and V. Syrmos. *Optimal Control*, 3rd ed. John Wiley & Sons, Inc., Hoboken, NJ, 2012.

[30] F. L. Lewis and K. G. Vamvoudakis. Reinforcement learning for partially observable dynamic processes: Adaptive dynamic programming using measured output data. *IEEE Transactions on Systems, Man, and Cybernetics, Part B: Cybernetics*, 41(1):14–25, 2011.

[31] F. L. Lewis and D. Vrabie. Reinforcement learning and adaptive dynamic programming for feedback control. *IEEE Circuits and Systems Magazine*, 9(3):32–50, 2009.

[32] H. Li, D. Liu, and D. Wang. Integral reinforcement learning for linear continuous-time zero-sum games with completely unknown dynamics. *IEEE Transactions on Automation Science and Engineering*, 11(3):706–714, July 2014.

[33] I. Mareels and J. W. Polderman. *Adaptive Systems: An Introduction.* Birkhäuser, 1996.

[34] H. Modares and F. Lewis. Linear quadratic tracking control of partially-unknown continuous-time systems using reinforcement learning. *IEEE Transactions on Automatic Control*, 59(11):3051–3056, 2014.

[35] H. Modares, F. L. Lewis, and M.-B. N. Sistani. Online solution of nonquadratic two-player zero-sum games arising in the H_∞ control of constrained input systems. *International Journal of Adaptive Control and Signal Processing*, 28(3–5):232–254, 2014.

[36] J. J. Murray, C. J. Cox, G. G. Lendaris, and R. Saeks. Adaptive dynamic programming. *IEEE Transactions on Systems, Man, and Cybernetics, Part C: Applications and Reviews*, 32(2):140–153, 2002.

[37] C. Qin, H. Zhang, and Y. Luo. Model-free H_∞ control design for unknown continuous-time linear system using adaptive dynamic programming. *Asian Journal of Control*, 18(3):1–10, 2016.

[38] R. Song, F. L. Lewis, Q. Wei, and H. Zhang. Off-policy actor-critic structure for optimal control of unknown systems with disturbances. *IEEE Transactions on Cybernetics*, 46(5):1041–1050, 2015.

[39] G. Tao. *Adaptive Control Design and Analysis*. John Wiley & Sons, 2003.

[40] K. G. Vamvoudakis, P. J. Antsaklis, W. E. Dixon, J. P. Hespanha, F. L. Lewis, H. Modares, and B. Kiumarsi. Autonomy and machine intelligence in complex systems: A tutorial. In: Proceedings of the American Control Conference (ACC'15), pp. 5062–5079, Chicago, IL, 2015.

[41] K. G. Vamvoudakis and F. L. Lewis. Multi-player non-zero-sum games: online adaptive learning solution of coupled Hamilton–Jacobi equations. *Automatica*, 47(8):1556–1569, 2011.

[42] D. Vrabie, O. Pastravanu, M. Abu-Khalaf, and F. Lewis. Adaptive optimal control for continuous-time linear systems based on policy iteration. *Automatica*, 45(2):477–484, 2009.

[43] F.-Y. Wang, H. Zhang, and D. Liu. Adaptive dynamic programming: An introduction. *IEEE Computational Intelligence Magazine*, 4(2):39–47, 2009.

[44] C. Watkins. Learning from delayed rewards. PhD Thesis, King's College of Cambridge, 1989.

[45] Q. Wei, H. Zhang, and J. Dai. Model-free multiobjective approximate dynamic programming for discrete-time nonlinear systems with general performance index functions. *Neurocomputing*, 72(7):1839–1848, 2009.

[46] P. J. Werbos. Neural networks for control and system identification. In: Proceedings of the 28th IEEE Conference on Decision and Control, pp. 260–265, Tampa, FL, 1989.

[47] P. J. Werbos. Intelligence in the brain: A theory of how it works and how to build it. *Neural Networks*, 22(3):200–212, 2009.

[48] H. Xu, S. Jagannathan, and F. L. Lewis. Stochastic optimal control of unknown linear networked control system in the presence of random delays and packet losses. *Automatica*, 48(6):1017–1030, 2012.

[49] H. Zhang, Q. Wei, and D. Liu. An iterative adaptive dynamic programming method for solving a class of nonlinear zero-sum differential games. *Automatica*, 47(1):207–214, 2011.

CHAPTER 3

SEMI-GLOBAL ADAPTIVE DYNAMIC PROGRAMMING

This chapter extends the results developed in the previous chapter to handle affine nonlinear systems via neural network-based approximation. The block diagram of such a setting is shown in Figure 3.1. An online learning method with convergence analysis is provided and it achieves *semi-global* stabilization for nonlinear systems in that the domain of attraction can be made arbitrarily large, but bounded, by tuning the controller parameters or design functions.

3.1 PROBLEM FORMULATION AND PRELIMINARIES

3.1.1 Problem Formulation

Consider a nonlinear system of the form

$$\dot{x} = f(x) + g(x)u \tag{3.1}$$

where $x \in \mathbb{R}^n$ is the state, $u \in \mathbb{R}^m$ is the control input, $f : \mathbb{R}^n \to \mathbb{R}^n$ and $g : \mathbb{R}^n \to \mathbb{R}^{n \times m}$ are locally Lipschitz mappings with $f(0) = 0$.

The objective is to find a control policy u that minimizes the following cost:

$$J(x_0; u) = \int_0^\infty r(x(t), u(t))dt, \quad x(0) = x_0 \tag{3.2}$$

Robust Adaptive Dynamic Programming, First Edition. Yu Jiang and Zhong-Ping Jiang.
© 2017 by The Institute of Electrical and Electronics Engineers, Inc. Published 2017 by John Wiley & Sons, Inc.

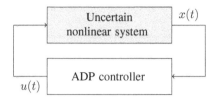

FIGURE 3.1   ADP-based online learning control for uncertain nonlinear systems.

where $r(x, u) = q(x) + u^T R(x)u$, with $q(x)$ a positive definite function, and $R(x)$ is symmetric and positive definite for all $x \in \mathbb{R}^n$. Notice that, the purpose of specifying $r(x, u)$ in this quadratic form is to guarantee that an optimal control policy can be explicitly determined, if it exists.

Before solving the optimal control problem, let us assume there exists at least an *admissible* control policy, as described in Assumption 3.1.1 below.

Assumption 3.1.1   *There exists a feedback control policy $u_0 : \mathbb{R}^n \to \mathbb{R}^m$ that globally asymptotically stabilizes the system (3.1) at the origin and the associated cost as defined in (3.2) is finite.*

In addition, we assume the existence of an optimal control policy. This assumption also implies the existence of an optimal value function, which is the solution to the well-known Hamilton–Jacobi–Bellman (HJB) equation. Before introducing this assumption, let us define C^1 as the set of all continuously differentiable functions, and $\mathcal{P}$ as the set of all functions in C^1 that are also positive definite and radially unbounded.

Assumption 3.1.2   *There exists $V^* \in \mathcal{P}$, such that the HJB equation holds*

$$\mathcal{H}(V^*) = 0 \tag{3.3}$$

where $\mathcal{H}$ is defined as

$$\mathcal{H}(V)(x) \triangleq \nabla V^T(x)f(x) + q(x) - \frac{1}{4}\nabla V^T(x)g(x)R^{-1}(x)g^T(x)\nabla V(x)$$

Under Assumption 3.1.2, it is easy to see that $V^*(x)$ is a well-defined Lyapunov function for the closed-loop system comprised of (3.1) and the control policy

$$u^*(x) = -\frac{1}{2}R^{-1}(x)g^T(x)\nabla V^*(x) \tag{3.4}$$

globally asymptotically stabilizes (3.1) at $x = 0$.

Then, as shown in [22], u^* is the optimal control policy, and $V^*(x_0)$ gives the optimal cost at the initial condition $x(0) = x_0$, that is,

$$V^*(x_0) = \min_u J(x_0; u) = J(x_0; u^*), \forall x_0 \in \mathbb{R}^n \tag{3.5}$$

It can also be shown that V^* is the unique solution to the HJB equation (3.3) with $V^* \in \mathcal{P}$. Indeed, let $\hat{V} \in \mathcal{P}$ be another solution to (3.3). Then, along the solutions of the closed-loop system composed of (3.1) and the control policy

$$u = \hat{u}(x) = -\frac{1}{2}R^{-1}(x)g^T(x)\nabla\hat{V}(x), \tag{3.6}$$

it follows that

$$\hat{V}(x_0) = V^*(x_0) - \int_0^\infty (u^* - \hat{u})^T R(x)(u^* - \hat{u})dt,$$
$$\leq V^*(x_0), \tag{3.7}$$

for any $x_0 \in \mathbb{R}^n$. Comparing (3.5) and (3.7), it can be concluded that

$$V^*(x) = \hat{V}(x), \quad \forall x \in \mathbb{R}^n \tag{3.8}$$

3.1.2 Conventional Policy Iteration

If there exists a class-$\mathcal{P}$ function which solves the HJB equation (3.3), an optimal control policy can be easily obtained. However, the nonlinear HJB equation (3.3) is very difficult to be solved analytically, in general. As a result, numerical methods are developed to approximate the solution. In particular, the following policy iteration method is widely used [21]. It can be reviewed as the nonlinear version of the Kleinman's algorithm [13].

Algorithm 3.1.3   *Conventional policy iteration algorithm*

(1) *Policy evaluation:*
   For $i = 0, 1, \dots$, solve for the cost function $V_i(x) \in C^1$, with $V_i(0) = 0$, from the following partial differential equation:

$$\nabla V_i^T(x)[f(x) + g(x)u_i] + r(x, u_i) = 0 \tag{3.9}$$

(2) *Policy improvement:*
   Update the control policy by

$$u_{i+1}(x) = -\frac{1}{2}R^{-1}(x)g^T(x)\nabla V_i(x) \tag{3.10}$$

The convergence property of the conventional policy iteration algorithm is given in the following theorem, which is an extension of [21, Theorem 4].

Theorem 3.1.4   *Suppose Assumptions 3.1.1 and 3.1.2 hold, and the solution $V_i(x) \in C^1$ satisfying (3.9) exists for $i = 0, 1, \cdots$. Let $V_i(x)$ and $u_{i+1}(x)$ be the functions obtained from (3.9) and (3.10). Then, the following properties hold, $\forall i = 0, 1, \ldots$*

    (1) *$V^*(x) \leq V_{i+1}(x) \leq V_i(x)$, $\forall x \in \mathbb{R}^n$.*

    (2) *u_i is globally stabilizing.*

    (3) *Let $\lim_{i \to \infty} V_i(x_0) = V(x_0)$ and $\lim_{i \to \infty} u_i(x_0) = u(x_0)$, $\forall x_0 \in \mathbb{R}^n$. Then, $V^* = V$ and $u^* = u$, if $V \in C^1$.*

Proof: See Appendix C.1. ∎

3.2 SEMI-GLOBAL ONLINE POLICY ITERATION

In this section, solutions to (3.9) and (3.10) are approximated using neural networks.

3.2.1 Off-Policy Learning

To begin with, consider the system

$$\dot{x} = f(x) + g(x)(u_0 + e) \tag{3.11}$$

where u_0 is the feedback control policy satisfying Assumption 3.1.1 and e is a bounded time-varying function, representing the exploration noise injected for learning purpose.

Due to the existence of the exploration noise, we need to ensure that the solutions of system (3.11) are forward complete, that is, are well defined for all positive time. For this purpose, we introduce the following assumption.

Assumption 3.2.1   *The closed-loop system (3.11) is input-to-state stable (ISS) when e, the exploration noise, is considered as the input.*

Under Assumption 3.2.1, (3.11) can be rewritten as

$$\dot{x} = f(x) + g(x)u_i(x) + g(x)v_i \tag{3.12}$$

where $v_i = u_0 - u_i + e$. For each $i \geq 0$, the time derivative of $V_i(x)$ along the solutions of (3.12) satisfies

$$\begin{aligned}
\dot{V}_i &= \nabla V_i^T(x)[f(x) + g(x)u_i + g(x)v_i] \\
&= -q(x) - u_i^T R(x)u_i + \nabla V_i^T(x)g(x)v_i \\
&= -q(x) - u_i^T R(x)u_i - 2u_{i+1}^T R(x)v_i
\end{aligned} \tag{3.13}$$

Integrating both sides of (3.13) on any time interval $[t, t + T]$, it follows that

$$V_i(x(t + T)) - V_i(x(t)) = - \int_t^{t+T} \left[q(x) + u_i^T R(x)u_i + 2u_{i+1}^T R(x)v_i \right] d\tau \quad (3.14)$$

Now, if an admissible control policy u_i is given, the unknown functions $V_i(x)$ and $u_{i+1}(x)$ can be approximated using (3.14). To be more specific, for any given compact set $\Omega \subset \mathbb{R}^n$ containing the origin as an interior point, let $\phi_j : \mathbb{R}^n \to \mathbb{R}$ and $\psi_j : \mathbb{R}^n \to \mathbb{R}^m$, with $j = 1, 2, \cdots$, be two infinite sequences of linearly independent smooth basis functions on Ω, vanishing at the origin. Then, by approximation theory [18], for each $i = 0, 1, \cdots$, the value function V_i and the control policy u_{i+1} can be approximated by

$$\hat{V}_i(x) = \sum_{j=1}^{N_1} \hat{c}_{i,j} \phi_j(x), \quad (3.15)$$

$$\hat{u}_{i+1}(x) = \sum_{j=1}^{N_2} \hat{w}_{i,j} \psi_j(x), \quad (3.16)$$

where $N_1 > 0$, $N_2 > 0$ are two sufficiently large integers, and $\hat{c}_{i,j}$, $\hat{w}_{i,j}$ are constant weights to be determined.

Replacing V_i, u_i, and u_{i+1} in (3.14) with their approximations (3.15) and (3.16), we obtain

$$\sum_{j=1}^{N_1} \hat{c}_{i,j} [\phi_j(x(t_{k+1})) - \phi_j(x(t_k))]$$

$$= - \int_{t_k}^{t_{k+1}} 2 \sum_{j=1}^{N_2} \hat{w}_{i,j} \psi_j^T(x) R(x) \hat{v}_i dt$$

$$- \int_{t_k}^{t_{k+1}} \left[q(x) + \hat{u}_i^T R(x) \hat{u}_i \right] dt + e_{i,k} \quad (3.17)$$

where $\hat{u}_0 = u_0$, $\hat{v}_i = u_0 + e - \hat{u}_i$, and $\{t_k\}_{k=0}^l$ is a strictly increasing sequence with $l > 0$ a sufficiently large integer. Then, the weights $\hat{c}_{i,j}$ and $\hat{w}_{i,j}$ can be solved in the sense of least-squares (i.e., by minimizing $\sum_{k=0}^l e_{i,k}^2$).

Now, starting from $u_0(x)$, two sequences $\{\hat{V}_i(x)\}_{i=0}^\infty$ and $\{\hat{u}_{i+1}(x)\}_{i=0}^\infty$ can be generated via the online policy iteration technique (3.17). Next, we show the convergence of the sequences to $V_i(x)$ and $u_{i+1}(x)$, respectively.

3.2.2 Convergence Analysis

Assumption 3.2.2 *There exist $l_0 > 0$ and $\delta > 0$, such that for all $l \geq l_0$, we have*

$$\frac{1}{l} \sum_{k=0}^l \theta_{i,k}^T \theta_{i,k} \geq \delta I_{N_1+N_2} \quad (3.18)$$

where

$$
\theta_{i,k}^T = \begin{bmatrix}
\phi_1(x(t_{k+1})) - \phi_1(x(t_k)) \\
\phi_2(x(t_{k+1})) - \phi_2(x(t_k)) \\
\vdots \\
\phi_{N_1}(x(t_{k+1})) - \phi_{N_1}(x(t_k)) \\
2 \int_{t_k}^{t_{k+1}} \psi_1^T(x)R(x)\hat{v}_i dt \\
2 \int_{t_k}^{t_{k+1}} \psi_2^T(x)R(x)\hat{v}_i dt \\
\vdots \\
2 \int_{t_k}^{t_{k+1}} \psi_{N_2}^T(x)R(x)\hat{v}_i dt
\end{bmatrix} \in \mathbb{R}^{N_1+N_2}
$$

Theorem 3.2.3 *Under Assumptions 3.2.1 and 3.2.2, for each $i \geq 0$ and given $\epsilon > 0$, there exist $N_1^* >$ and $N_2^* > 0$, such that*

$$
\left| \sum_{j=1}^{N_1} \hat{c}_{i,j}\phi_j(x) - V_i(x) \right| < \epsilon \tag{3.19}
$$

$$
\left| \sum_{j=1}^{N_2} \hat{w}_{i,j}\psi_j(x) - u_{i+1}(x) \right| < \epsilon \tag{3.20}
$$

for all $x \in \Omega$, if $N_1 > N_1^*$ and $N_2 > N_2^*$. ∎

Proof: See the Appendix.

Corollary 3.2.4 *Assume $V^*(x)$ and $u^*(x)$ exist. Then, under Assumptions 3.2.1 and 3.2.2, for any arbitrary $\epsilon > 0$, there exist integers $i^* > 0$, $N_1^{**} > 0$, and $N_2^{**} > 0$, such that*

$$
\left| \sum_{j=1}^{N_1} \hat{c}_{i^*,j}\phi_j(x) - V^*(x) \right| \leq \epsilon \tag{3.21}
$$

$$
\left| \sum_{j=1}^{N_2} \hat{w}_{i^*,j}\psi_j(x) - u^*(x) \right| \leq \epsilon \tag{3.22}
$$

for all $x \in \Omega$, if $N_1 > N_1^{**}$ and $N_2 > N_2^{**}$.

Proof: By Theorem 3.1.4, there exists $i^* > 0$, such that

$$
|V_{i^*}(x) - V^*(x)| \leq \frac{\epsilon}{2} \tag{3.23}
$$

$$
|u_{i^*+1}(x) - u^*(x)| \leq \frac{\epsilon}{2}, \forall x \in \Omega \tag{3.24}
$$

By Theorem 3.2.3, there exist $N_1^{**} > 0$ and $N_2^{**} > 0$, such that for all $N_1 > N_1^{**}$ and $N_2 > N_2^{**}$,

$$\left| \sum_{j=1}^{N_1} \hat{c}_{i*,j} \phi_j(x) - V_{i*}(x) \right| \leq \frac{\epsilon}{2} \tag{3.25}$$

$$\left| \sum_{j=1}^{N_2} \hat{w}_{i*,j} \psi_j(x) - u_{i*+1}(x) \right| \leq \frac{\epsilon}{2}, \forall x \in \Omega \tag{3.26}$$

The corollary is thus proved by using the triangle inequality [20]. ∎

3.2.3 Stability Analysis

Before applying the proposed ADP algorithm, let Ω be the compact set over which neural network-based approximation is carried out. In general, neural networks are not able to approximate well all nonlinear functions defined on the entire state-space $\mathbb{R}^n$, except on compact sets. On the other hand, after the convergence of the learning algorithm is attained, the newly learned control policy may not be immediately applicable because there is a possibility that system solutions can go out of the set Ω, causing instability issues. Therefore, it is of importance to quantify the region of attraction of the closed-loop system with the new control policy. To this end, let us define the following set

$$\Omega_i = \{x | V_i(x) \leq d_\Omega\} \tag{3.27}$$

where $d_\Omega > 0$ is a constant such that

$$V_i(x) \leq d_\Omega \Rightarrow x \in \Omega \tag{3.28}$$

Theorem 3.2.5   *Under Assumptions 3.1.2, 3.2.1, and 3.2.2, the closed-loop system comprised of (3.1) and $u = \hat{u}_{i+1}$ is asymptotically stable at the origin, if*

$$q(x) > \left(u_{i+1} - \hat{u}_{i+1} \right)^T R(x) \left(u_{i+1} - \hat{u}_{i+1} \right), \quad \forall x \in \Omega \backslash \{0\} \tag{3.29}$$

In addition, let

$$\hat{\Omega}_i = \{x | \hat{V}_i(x) \leq d_\Omega - \epsilon\} \tag{3.30}$$

with $\epsilon > 0$ an arbitrarily small constant. Then, there exist $i > 0$, $N_1^{**} > 0$, and $N_2^{**} > 0$, such that the following implication holds, if $N_1 > N_1^{**}$ and $N_2 > N_1^{**}$,

$$x(0) \in \hat{\Omega}_i \Rightarrow x(t) \in \Omega_i, \forall t \geq 0 \tag{3.31}$$

Proof: Along the solutions of (3.1) with $u = \hat{u}_{i+1}$, we have

$$
\begin{aligned}
\dot{V}_i &= \nabla V_i^T(x)(f(x) + g(x)\hat{u}_{i+1}) \\
&= \nabla V_i^T(x)(f(x) + g(x)u_{i+1}) + \nabla V_i^T(x)g(x)(\hat{u}_{i+1} - u_{i+1}) \\
&= -q(x) - u_{i+1}^T R(x)u_{i+1} - (u_{i+1} - u_i)^T R(x)(u_{i+1} - u_i) \\
&\quad -2u_{i+1}^T R(x)(\hat{u}_{i+1} - u_{i+1}) \\
&\leq -q(x) + u_{i+1}^T R(x)u_{i+1} - 2u_{i+1}^T R(x)\hat{u}_{i+1} \\
&\leq -q(x) + \left(u_{i+1} - \hat{u}_{i+1}\right)^T R(x)\left(u_{i+1} - \hat{u}_{i+1}\right)
\end{aligned}
$$

for all $x \in \Omega$ and $x \neq 0$. Hence, Ω_i is an invariant set of the closed-loop system.

Furthermore, by Corollary 3.2.4, we know that there exist $i > 0$, $N_1^{**} > 0$, and $N_2^{**} > 0$, such that

$$
\hat{\Omega}_i \subseteq \Omega_i \tag{3.32}
$$

if $N_1 > N_1^{**}$ and $N_2 > N_1^{**}$. The proof is thus complete.   ∎

Remark 3.2.6   *The resulting control policy $\hat{u}_{i+1}$ achieves semi-global stabilization, in the sense that the origin of the closed-loop system composed of (3.1) and $\hat{u}_{i+1}$ is asymptotically stable and $\hat{u}_{i+1}$ can be designed such that any given compact set can be included in the region of attraction [12].*

3.2.4   Implementation Issues

The semi-global ADP algorithm is given in Algorithm 3.2.7, and a graphical illustration is provided in Figure 3.2.

Algorithm 3.2.7   *Off-policy ADP algorithm for nonlinear systems*

(1) *Initialization:*
   Determine the compact set $\Omega \in \mathbb{R}^n$ for approximation. Find an initial control policy u_0 and let $i \leftarrow 0$.

(2) *Online data collection:*
   Apply the initial control policy $u = u_0 + e$ and collect the system state and input information.

(3) *Policy evaluation and improvement:*
   Solve $\hat{c}_{i,j}$ and $\hat{w}_{i,j}$ from (3.17).

(4) *Stopping criterion:*
   Let $i \leftarrow i + 1$, and go to Step (2), until

$$
\sum_{j=1}^{N_1} |\hat{c}_{i,j} - \hat{c}_{i-1,j}|^2 \leq \epsilon_1 \tag{3.33}
$$

where the constant $\epsilon_1 > 0$ is a sufficiently small predefined threshold.

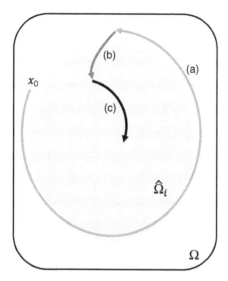

FIGURE 3.2   Illustration of the semi-global ADP algorithm. Ω is the compact set for neural network-based approximation. $\hat{\Omega}_i$ is an estimate of the region of attraction after learning. (a) Online data are collected and off-policy iteration is applied. (b) Exploration is terminated. (c) The actual control policy is improved.

(5) *Actual control policy improvement:*
   Terminate the exploration noise e and use $u = u_0$ as the control input. Once $x(t) \in \hat{\Omega}_i$, apply the approximate optimal control policy $u = \hat{u}_{i+1}$.

3.3 APPLICATION

Consider a car suspension system described by the following ordinary differential equations:

$$\dot{x}_1 = x_2 \tag{3.34}$$

$$\dot{x}_2 = -\frac{k_s(x_1-x_3) + k_n(x_1-x_3)^3 + b_s(x_2-x_4) - cu}{m_b} \tag{3.35}$$

$$\dot{x}_3 = x_3 \tag{3.36}$$

$$\dot{x}_4 = \frac{k_s(x_1-x_3) + k_n(x_1-x_3)^3 + b_s(x_2-x_4) - k_t x_3 - cu}{m_w} \tag{3.37}$$

where x_1, x_2, and m_b denote respectively the position, velocity, and mass of the car body; x_3, x_4, and m_w represent respectively the position, velocity, and mass of the wheel assembly; k_t, k_s, k_n, and b_s are the tyre stiffness, the linear suspension stiffness, the nonlinear suspension stiffness, and the damping ratio of the suspension; c is a constant relating the control signal to input force.

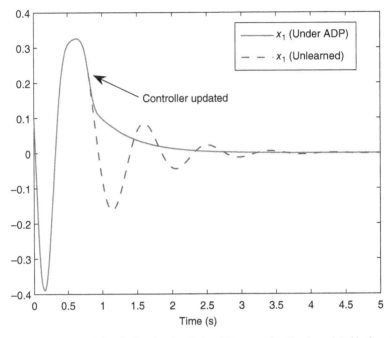

FIGURE 3.3   Base position during the simulation. The control policy is updated before $t = 1$.

It is assumed that $m_b \in [250, 350]$, $m_w \in [55, 65]$, $b_s \in [900, 1100]$, $k_s \in [15000, 17000]$, $k_n = k_s/10$, $k_t \in [180000, 200000]$, and $c > 0$. Then, as it can be easily checked, the uncontrolled system is globally asymptotically stable at the origin. To optimize the system performance, we are interested in reducing the following performance index

$$J(x_0; u) = \int_0^\infty \left(\sum_{i=1}^4 x_i^2 + u^2 \right) dt \qquad (3.38)$$

on the set

$$\Omega = \{x | x \in \mathbb{R}^4 \text{ and } |x_1| \leq 0.5, |x_2| \leq 5, |x_3| \leq 0.2, |x_4| \leq 10\}$$

The exploration noise is set to be the sum of sinusoidal signals with six different frequencies. The approximate optimal control policy is obtained after 19 iterations and the control policy is updated in less than one second after the algorithm started. The trajectories of the base position x_1 are shown in Figure 3.3. The initial value function and the improved one are compared in Figure 3.4. An estimate of the region of attraction associated with the approximate optimal control is illustrated in Figure 3.5.

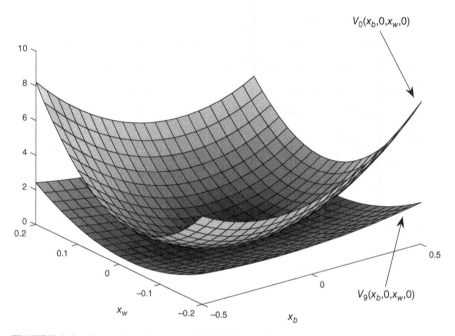

FIGURE 3.4 Comparison between the initial value function and the one improved after 9 iterations.

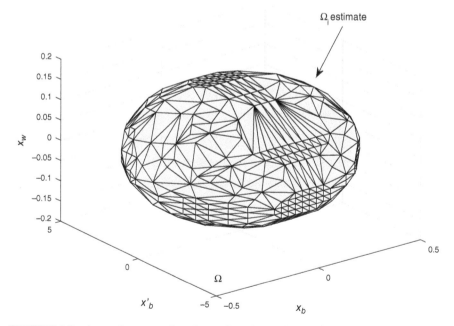

FIGURE 3.5 Approximate post-learning region of attraction projected to the space spanned by x_b, $\dot{x}_b$, and x_w.

3.4 NOTES

Two most frequently used techniques in reinforcement learning are value iteration and policy iteration. They are proposed in [3] and [7] for Markov Decision Processes. For continuous-time nonlinear dynamic systems, the policy iteration algorithm was developed in [14] and [21]. In all the iteration schemes, it is essential to approximate the value function in each iteration step [1, 2, 4, 15, 24]. When the system dynamics are uncertain, the approximation can be realized using online information via reinforcement learning and ADP methods [6, 17].

The first exact online policy iteration method for continuous-time affine nonlinear systems was proposed in [23], which requires partial system knowledge to be known exactly. This restriction has been removed in [9], which is a nonlinear extension of the ADP method for linear systems [8].

Neural network-based ADP methods for nonlinear control systems are being actively developed by a good number of researchers. Some recent theoretical results include ADP for non-affine nonlinear systems [5], ADP for saturated control design [25], ADP for nonlinear games [16], and ADP for nonlinear tracking problems [11, 19]. Unfortunately, there are three main limitations in neural network-based ADP design. First, the selection of the neurons is not trivial, especially when the system dynamic is uncertain. Second, the region of attraction changes if the control policy is updated. This can affect the stability of the closed-loop system. Third, global asymptotic stability usually cannot be attained by means of neural network-based control schemes. These issues will be tackled in Chapter 4. The extension proposed for the first time along this line is known under the name of global adaptive dynamic programming [10].

REFERENCES

[1] M. Abu-Khalaf and F. L. Lewis. Nearly optimal control laws for nonlinear systems with saturating actuators using a neural network HJB approach. *Automatica*, 41(5):779–791, 2005.

[2] R. W. Beard, G. N. Saridis, and J. T. Wen. Galerkin approximations of the generalized Hamilton-Jacobi-Bellman equation. *Automatica*, 33(12):2159–2177, 1997.

[3] R. E. Bellman. *Dynamic Programming*. Princeton University Press, Princeton, NJ, 1957.

[4] D. P. Bertsekas. *Dynamic Programming and Optimal Control*, 4th ed. Athena Scientific, Belmont, MA, 2007.

[5] T. Bian, Y. Jiang, and Z. P. Jiang. Adaptive dynamic programming and optimal control of nonlinear nonaffine systems. *Automatica*, 50(10):2624–2632, October 2014.

[6] K. Doya. Reinforcement learning in continuous time and space. *Neural Computation*, 12(1):219–245, 2000.

[7] R. Howard. *Dynamic Programming and Markov Processes*. MIT Press, Cambridge, MA, 1960.

[8] Y. Jiang and Z. P. Jiang. Computational adaptive optimal control for continuous-time linear systems with completely unknown dynamics. *Automatica*, 48(10):2699–2704, 2012.

[9] Y. Jiang and Z. P. Jiang. Robust adaptive dynamic programming and feedback stabilization of nonlinear systems. *IEEE Transactions on Neural Networks and Learning Systems*, 25(5):882–893, 2014.

[10] Y. Jiang and Z. P. Jiang. Global adaptive dynamic programming for continuous-time nonlinear systems. *IEEE Transactions on Automatic Control*, 60(11):2917–2929, 2015.

[11] R. Kamalapurkar, H. Dinh, S. Bhasin, and W. E. Dixon. Approximate optimal trajectory tracking for continuous-time nonlinear systems. *Automatica*, 51:40–48, 2015.

[12] H. K. Khalil. *Nonlinear Systems*, 3rd ed. Prentice Hall, Upper Saddle River, NJ, 2002.

[13] D. Kleinman. On an iterative technique for Riccati equation computations. *IEEE Transactions on Automatic Control*, 13(1):114–115, 1968.

[14] R. Leake and R.-W. Liu. Construction of suboptimal control sequences. *SIAM Journal on Control*, 5(1):54–63, 1967.

[15] F. L. Lewis and D. Vrabie. Reinforcement learning and adaptive dynamic programming for feedback control. *IEEE Circuits and Systems Magazine*, 9(3):32–50, 2009.

[16] D. Liu, H. Li, and D. Wang. Neural-network-based zero-sum game for discrete-time nonlinear systems via iterative adaptive dynamic programming algorithm. *Neurocomputing*, 110:92–100, 2013.

[17] J. J. Murray, C. J. Cox, G. G. Lendaris, and R. Saeks. Adaptive dynamic programming. *IEEE Transactions on Systems, Man, and Cybernetics, Part C: Applications and Reviews*, 32(2):140–153, 2002.

[18] M. J. D. Powell. *Approximation Theory and Methods*. Cambridge university press, 1981.

[19] C. Qin, H. Zhang, and Y. Luo. Online optimal tracking control of continuous-time linear systems with unknown dynamics by using adaptive dynamic programming. *International Journal of Control*, 87(5):1000–1009, 2014.

[20] W. Rudin. *Principles of Mathematical Analysis*, 3rd ed. McGraw-Hill, New York, NY, 1976.

[21] G. N. Saridis and C.-S. G. Lee. An approximation theory of optimal control for trainable manipulators. *IEEE Transactions on Systems, Man and Cybernetics*, 9(3):152–159, 1979.

[22] R. Sepulchre, M. Jankovic, and P. Kokotovic. *Constructive Nonlinear Control*. Springer-Verlag, New York, 1997.

[23] D. Vrabie and F. L. Lewis. Neural network approach to continuous-time direct adaptive optimal control for partially unknown nonlinear systems. *Neural Networks*, 22(3):237–246, 2009.

[24] F.-Y. Wang, H. Zhang, and D. Liu. Adaptive dynamic programming: An introduction. *IEEE Computational Intelligence Magazine*, 4(2):39–47, 2009.

[25] X. Yang, D. Liu, and D. Wang. Reinforcement learning for adaptive optimal control of unknown continuous-time nonlinear systems with input constraints. *International Journal of Control*, 87(3):553–566, 2014.

CHAPTER 4

GLOBAL ADAPTIVE DYNAMIC PROGRAMMING FOR NONLINEAR POLYNOMIAL SYSTEMS

This chapter brings more advanced optimization techniques, such as semidefinite programming (SDP) and sum-of-squares (SOS) programming, into ADP design (see Figure 4.1). The goal is to achieve adaptive suboptimal online learning and, at the same time, maintain global asymptotic stability of the closed-loop system.

4.1 PROBLEM FORMULATION AND PRELIMINARIES

4.1.1 Problem Formulation

Consider a nonlinear system described as follows:

$$\dot{x} = f(x) + g(x)u \tag{4.1}$$

where $x \in \mathbb{R}^n$ is the system state, $u \in \mathbb{R}^m$ is the control input, $f : \mathbb{R}^n \to \mathbb{R}^n$ and $g : \mathbb{R}^n \to \mathbb{R}^{n \times m}$ are polynomial mappings with $f(0) = 0$.

Similar to Chapter 3 and [17], we are interested in finding a control policy u that minimizes the following cost

$$J(x_0; u) = \int_0^\infty r(x(t), u(t))dt, \ x(0) = x_0 \tag{4.2}$$

Robust Adaptive Dynamic Programming, First Edition. Yu Jiang and Zhong-Ping Jiang.
© 2017 by The Institute of Electrical and Electronics Engineers, Inc. Published 2017 by John Wiley & Sons, Inc.

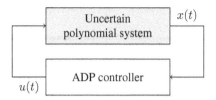

FIGURE 4.1   ADP-based online learning control for uncertain polynomial systems.

where $r(x, u) = q(x) + u^T R(x)u$, with $q(x)$ a positive definite polynomial function, and $R(x)$ a symmetric positive definite matrix of polynomials. Again, the purpose of specifying $r(x, u)$ in this form is to guarantee that an optimal control policy can be explicitly determined, if it exists.

In addition, similar as in Assumption 3.1.1, it is assumed that the optimization problem (4.2) has at least one feasible solution, as described below.

Assumption 4.1.1   *Consider system (4.1). There exist a function $V_0 \in \mathcal{P}$ and a feedback control policy $u_1 : \mathbb{R}^n \to \mathbb{R}^m$, such that*

$$\mathcal{L}(V_0(x), u_1(x)) \geq 0, \ \forall x \in \mathbb{R}^n \tag{4.3}$$

where, for any $V \in C^1$ and $u \in \mathbb{R}^m$, $\mathcal{L}(\cdot, \cdot)$ is defined as

$$\mathcal{L}(V, u) \triangleq -\nabla V^T(x)(f(x) + g(x)u) - r(x, u) \tag{4.4}$$

Under Assumption 4.1.1, the closed-loop system composed of (4.1) and $u = u_1(x)$ is globally asymptotically stable at the origin, with a well-defined Lyapunov function V_0. With this property, u_1 is admissible, since it not only globally asymptotically stabilizes the system (4.1) at the origin, but also assures that the cost $J(x_0; u_1)$ is finite, $\forall x_0 \in \mathbb{R}^n$. Indeed, integrating both sides of (4.3) along the trajectories of the closed-loop system composed of (4.1) and $u = u_1(x)$ on the interval $[0, +\infty)$, it is easy to show that

$$J(x_0; u_1) \leq V_0(x_0), \ \forall x_0 \in \mathbb{R}^n \tag{4.5}$$

Similar to Chapter 3, we assume the existence of the solution to the associated HJB by imposing the following assumption.

Assumption 4.1.2   *There exists $V^* \in \mathcal{P}$, such that the HJB equation holds*

$$\mathcal{H}(V^*) = 0 \tag{4.6}$$

where $\mathcal{H}$ is defined as

$$\mathcal{H}(V)(x) \triangleq \nabla V^T(x)f(x) + q(x) - \frac{1}{4}\nabla V^T(x)g(x)R^{-1}(x)g^T(x)\nabla V(x)$$

Under Assumption 4.1.2,

$$u^*(x) = -\frac{1}{2}R^{-1}(x)g^T(x)\nabla V^*(x) \tag{4.7}$$

is the optimal control policy, and the value function $V^*(x_0)$ gives the optimal cost at the initial condition $x(0) = x_0$ [28, Theorem 3.19]. As a result, the closed-loop system (4.1) and (4.7) is globally asymptotically stable at $x = 0$ with V^* as the Lyapunov function (see [14]). Finally, V^* is the unique solution to the HJB (4.6), as explained in Chapter 3.

4.1.2 Conventional Policy Iteration

Clearly, the conventional policy iteration algorithm (Algorithm 3.1.3) introduced in [24] is also applicable to the optimal control problem formulated above. For the readers' convenience, the algorithm is re-stated as follows.

Algorithm 4.1.3   *Conventional policy iteration algorithm*

(1) *Policy evaluation:*
   For $i = 1, 2, ...$, solve for the cost function $V_i(x) \in C^1$, with $V_i(0) = 0$, from the following partial differential equation.

$$\mathcal{L}(V_i(x), u_i(x)) = 0 \tag{4.8}$$

(2) *Policy improvement:*
   Update the control policy by

$$u_{i+1}(x) = -\frac{1}{2}R^{-1}(x)g^T(x)\nabla V_i(x) \tag{4.9}$$

The convergence property, summarized in the following corollary, is directly implied by Theorem 3.1.4.

Corollary 4.1.4   *Suppose Assumptions 4.1.1 and 4.1.2 hold, and the solution $V_i(x) \in C^1$ satisfying (4.8) exists, for $i = 1, 2, ...$ Let $V_i(x)$ and $u_{i+1}(x)$ be the functions obtained from (4.8) and (4.9), respectively. Then, the following properties hold, $\forall i = 0, 1, ...$*

(1) *$V^*(x) \leq V_{i+1}(x) \leq V_i(x)$, $\forall x \in \mathbb{R}^n$.*
(2) *u_{i+1} is globally stabilizing.*
(3) *Let $\lim_{i\to\infty} V_i(x_0) = V^*(x_0)$ and $\lim_{i\to\infty} u_i(x_0) = u^*(x_0)$, for any given x_0. Then, $V = V^*$ and $u = u^*$, if $V^* \in C^1$.*

Finding the exact solution to (4.8) is still a non-trivial problem. In Chapter 3, this solution is approximated using neural networks. Indeed, many related approximation methods have been extensively studied in recent years (see, e.g., [12, 33], and [2]). However, most of these methods fail to achieve global stabilization. In addition, in order to reduce the approximation error, huge computational complexity is almost inevitable. To address these issues, the idea of relaxation will be applied in Section 4.2.

4.2 RELAXED HJB EQUATION AND SUBOPTIMAL CONTROL

In this section, we consider an auxiliary optimization problem, which allows us to obtain a suboptimal solution to the minimization problem (4.2) subject to (4.1). For simplicity, we will omit the arguments of functions whenever there is no confusion in the context.

Problem 4.2.1 (Relaxed optimal control problem)

$$\min_{V} \quad \int_{\Omega} V(x)dx \tag{4.10}$$

$$\text{s.t.} \quad \mathcal{H}(V) \leq 0 \tag{4.11}$$

$$V \in \mathcal{P} \tag{4.12}$$

where $\Omega \subset \mathbb{R}^n$ is an arbitrary compact set containing the origin as an interior point. As a subset of the state space, Ω describes the area in which the system performance is expected to be improved the most.

Remark 4.2.2   *Notice that Problem 4.2.1 is called a* relaxed *problem of (4.6). Indeed, if we restrict this problem by replacing the inequality constraint (4.11) with the equality constraint (4.6), there will be only one feasible solution left and the objective function can thus be neglected. As a result, Problem 4.2.1 reduces to the problem of solving (4.6).*

After relaxation, some useful facts about Problem 4.2.1 are summarized as follows.

Theorem 4.2.3   *Under Assumptions 4.1.1 and 4.1.2, the following hold.*

(1) *Problem 4.2.1 has a nonempty feasible set.*
(2) *Let V be a feasible solution to Problem 4.2.1. Then, the control policy*

$$\bar{u}(x) = -\frac{1}{2}R^{-1}g^T \nabla V \tag{4.13}$$

is globally stabilizing.
(3) *For any $x_0 \in \mathbb{R}^n$, an upper bound of the cost of the closed-loop system comprised of (4.1) and (4.13) is given by $V(x_0)$, that is,*

$$J(x_0; \bar{u}) \leq V(x_0) \tag{4.14}$$

(4) *Along the trajectories of the closed-loop system (4.1) and (4.7), the following inequalities hold, for any $x_0 \in \mathbb{R}^n$:*

$$V(x_0) + \int_0^\infty \mathcal{H}(V(x(t)))dt \leq V^*(x_0) \leq V(x_0) \tag{4.15}$$

(5) V^*, *as introduced in Assumption 4.1.2, is a global optimal solution to Problem 4.2.1.*

Proof: (1) Define $u_0 = -\frac{1}{2}R^{-1}g^T\nabla V_0$. Then,

$$
\begin{aligned}
\mathcal{H}(V_0) &= \nabla V_0^T(f + gu_0) + r(x, u_0)\\
&= \nabla V_0^T(f + gu_1) + r(x, u_1)\\
&\quad + \nabla V_0^T g(u_0 - u_1) + u_0^T Ru_0 - u_1^T Ru_1\\
&= \nabla V_0^T(f + gu_1) + r(x, u_1)\\
&\quad - 2u_0^T Ru_0 - 2u_0^T Ru_1 + u_0^T Ru_0 - u_1^T Ru_1\\
&= \nabla V_0^T(f + gu_1) + r(x, u_1) - (u_0 - u_1)^T R(u_0 - u_1)\\
&\leq 0
\end{aligned}
$$

Hence, V_0 is a feasible solution to Problem 4.2.1.

(2) To show global asymptotic stability, we only need to prove that V is a well-defined Lyapunov function for the closed-loop system composed of (4.1) and (4.13). Indeed, along the solutions of the closed-loop system, it follows that

$$
\begin{aligned}
\dot{V} &= \nabla V^T(f + g\bar{u})\\
&= \mathcal{H}(V) - r(x, \bar{u})\\
&\leq -q(x)
\end{aligned}
$$

Therefore, the system is globally asymptotically stable at the origin [14].

(3) Along the solutions of the closed-loop system comprised of (4.1) and (4.13), we have

$$
\begin{aligned}
V(x_0) &= -\int_0^T \nabla V^T(f + g\bar{u})dt + V(x(T))\\
&= \int_0^T [r(x, \bar{u}) - \mathcal{H}(V)]dt + V(x(T))\\
&\geq \int_0^T r(x, \bar{u})dt + V(x(T)) \tag{4.16}
\end{aligned}
$$

By (2), $\lim_{T \to +\infty} V(x(T)) = 0$. Therefore, letting $T \to +\infty$, by (4.16) and (4.2), we have

$$V(x_0) \geq J(x_0, \bar{u}) \tag{4.17}$$

(4) By (3), we have

$$V(x_0) \geq J(x_0; \bar{u}) \geq \min_u J(x_0; \bar{u}) = V^*(x_0) \tag{4.18}$$

Hence, the second inequality in (4.15) is proved.

On the other hand,

$$
\begin{aligned}
\mathcal{H}(V) &= \mathcal{H}(V) - \mathcal{H}(V^*) \\
&= (\nabla V - \nabla V^*)^T (f + gu^*) + r(x, \bar{u}) \\
&\quad + \nabla V^T g(\bar{u} - u^*) - r(x, u^*) \\
&= (\nabla V - \nabla V^*)^T (f + gu^*) - (\bar{u} - u^*)^T R(\bar{u} - u^*) \\
&\leq (\nabla V - \nabla V^*)^T (f + gu^*)
\end{aligned}
$$

Integrating the above equation along the solutions of the closed-loop system (4.1) and (4.7) on the interval $[0, +\infty)$, we have

$$V(x_0) - V^*(x_0) \leq - \int_0^\infty \mathcal{H}(V(x))dt \tag{4.19}$$

(5) By (3), for any feasible solution V to Problem 4.2.1, we have $V^*(x) \leq V(x)$. Hence,

$$\int_\Omega V^*(x)dx \leq \int_\Omega V(x)dx \tag{4.20}$$

which implies that V^* is a global optimal solution.   ∎

The proof is therefore complete.

Remark 4.2.4   *A feasible solution V to Problem 4.2.1 may not necessarily provide the true cost associated with the control policy $\bar{u}$ defined in (4.13). However, by Theorem 4.2.3, we see V can be viewed as an upper bound or an overestimate of the actual cost, inspired by the concept of underestimator in [34]. Further, V serves as a Lyapunov function for the closed-loop system and can be more easily parameterized than the actual value function. For simplicity, V is still called the value function in the remainder of this chapter.*

4.3 SOS-BASED POLICY ITERATION FOR POLYNOMIAL SYSTEMS

The inequality constraint (4.11) introduced in Problem 4.2.1 brings the freedom of specifying desired analytic forms of the value function. However, solving (4.11) is non-trivial in general, even for polynomial systems (see, e.g., [4, 6, 19, 29, 36]). To overcome this difficulty, this section develops a policy iteration method for polynomial systems using SOS-based methods [3, 21]. See Appendix B for more details on SOS.

4.3.1 Polynomial Parametrization

Let us first assume $R(x)$ is a constant, real symmetric matrix. Notice that $\mathcal{L}(V_i, u_i)$ is a polynomial in x, if both u_i and V_i are polynomials in x. Then, the following implication holds

$$\mathcal{L}(V_i, u_i) \text{ is SOS} \Rightarrow \mathcal{L}(V_i, u_i) \geq 0 \tag{4.21}$$

In addition, for computational simplicity, we would like to find some positive integer r, such that $V_i \in \mathbb{R}[x]_{2,2r}$. Then, the new control policy u_{i+1} obtained from (4.25) is a vector of polynomials.

Based on the above discussion, the following assumption is given to replace Assumption 4.1.1.

Assumption 4.3.1 *There exist smooth mappings $V_0 : \mathbb{R}^n \to \mathbb{R}$ and $u_1 : \mathbb{R}^n \to \mathbb{R}^m$, such that $V_0 \in \mathbb{R}[x]_{2,2r} \cap \mathcal{P}$ and $\mathcal{L}(V_0, u_1)$ is SOS.*

4.3.2 SOS-Based Policy Iteration

Now, we are ready to propose a relaxed policy iteration scheme. Similar as in other policy iteration-based iterative schemes, an initial globally stabilizing (and admissible) control policy has been assumed in Assumption 4.3.1.

Algorithm 4.3.2 *SOS-based policy iteration algorithm*

(1) *Policy evaluation:*
For $i = 1, 2, \ldots$, solve for an optimal solution p_i to the following optimization problem:

$$\min_{p} \int_{\Omega} V(x)dx \tag{4.22}$$

$$\text{s.t. } \mathcal{L}(V, u_i) \quad \text{is} \quad \text{SOS} \tag{4.23}$$

$$V_{i-1} - V \quad \text{is} \quad \text{SOS} \tag{4.24}$$

where $V = p^T[x]_{2,2r}$. Then, define $V_i = p_i^T[x]_{2,2r}$.

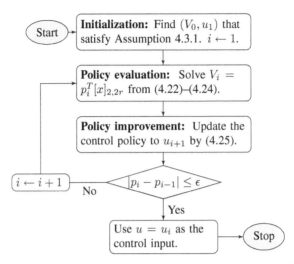

FIGURE 4.2   Flowchart of the SOS-based policy iteration. *Source*: Jiang, 2015. Reproduced with permission of IEEE.

(2) *Policy improvement:*
   Update the control policy by

$$u_{i+1} = -\frac{1}{2}R^{-1}g^T\nabla V_i \tag{4.25}$$

Then, go to Step (1) with i replaced by i + 1.

A flowchart of the practical implementation is given in Figure 4.2.

Remark 4.3.3   *The optimization problem (4.22)–(4.24) is a well-defined SOS program. Indeed, the objective function (4.22) is linear in p, since for any $V = p^T[x]_{2,2r}$, we have $\int_\Omega V(x)dx = c^Tp$, with $c = \int_\Omega [x]_{2,2r}dx$.*

Theorem 4.3.4   *Under Assumptions 4.1.2 and 4.3.1, the following statements are true, for $i = 1, 2, \ldots$*

(1) *The SOS program (4.22)–(4.24) has a nonempty feasible set.*
(2) *The closed-loop system comprised of (4.1) and $u = u_i$ is globally asymptotically stable at the origin.*
(3) *$V_i \in \mathcal{P}$. In particular, the following inequalities hold.*

$$V^*(x_0) \le V_i(x_0) \le V_{i-1}(x_0), \quad \forall x_0 \in \mathbb{R}^n \tag{4.26}$$

(4) *There exists $V_\infty(x)$ satisfying $V_\infty(x) \in \mathbb{R}[x]_{2,2r} \cap P$, such that, for any $x_0 \in \mathbb{R}^n$, $\lim_{i \to \infty} V_i(x_0) = V_\infty(x_0)$.*

(5) *Along the solutions of the system composed of (4.1) and (4.7) the following inequalities hold.*

$$0 \leq V_\infty(x_0) - V^*(x_0) \leq - \int_0^\infty \mathcal{H}(V_\infty(x(t)))dt \tag{4.27}$$

Proof: (1) Let us prove it by mathematical induction.

(i) Suppose $i = 1$, under Assumption 4.3.1, we know $\mathcal{L}(V_0, u_1)$ is SOS. Hence, $V = V_0$ is a feasible solution to the problem (4.22)–(4.24).

(ii) Let $V = V_{j-1}$ be an optimal solution to the problem (4.22)–(4.24) with $i = j - 1 > 1$. We show that $V = V_{j-1}$ is a feasible solution to the same problem with $i = j$.

Indeed, by definition,

$$u_j = -\frac{1}{2} R^{-1} g^T \nabla V_{j-1},$$

and

$$\begin{aligned}
\mathcal{L}(V_{j-1}, u_j) &= -\nabla V_{j-1}^T (f + g u_j) - r(x, u_j) \\
&= \mathcal{L}(V_{j-1}, u_{j-1}) - \nabla V_{j-1}^T g(u_j - u_{j-1}) \\
&\quad + u_{j-1}^T R u_{j-1} - u_j^T R u_j \\
&= \mathcal{L}(V_{j-1}, u_{j-1}) + (u_j - u_{j-1})^T R(u_j - u_{j-1})^T
\end{aligned}$$

Under the induction assumption, we know $V_{j-1} \in \mathbb{R}[x]_{2,2r}$ and $\mathcal{L}(V_{j-1}, u_{j-1})$ are SOS. Hence, $\mathcal{L}(V_{j-1}, u_j)$ is SOS. As a result, V_{j-1} is a feasible solution to the SOS program (4.22)–(4.24) with $i = j$.

(2) Again, we prove by induction.

(i) Suppose $i = 1$, under Assumption 4.3.1, u_1 is globally stabilizing. Also, we can show that $V_1 \in P$. Indeed, for each $x_0 \in \mathcal{R}^n$ with $x_0 \neq 0$, we have

$$V_1(x_0) \geq \int_0^\infty r(x, u_1)dt > 0 \tag{4.28}$$

By (4.28) and the constraint (4.24), under Assumption 4.1.2, it follows that

$$V^* \leq V_1 \leq V_0 \tag{4.29}$$

Since both V^* and V_0 are assumed to belong to P, we conclude that $V_1 \in P$.

(ii) Suppose u_{i-1} is globally stabilizing, and $V_{i-1} \in P$ for $i > 1$. Let us show that u_i is globally stabilizing, and $V_i \in P$.

Indeed, along the solutions of the closed-loop system composed of (4.1) and $u = u_i$, it follows that

$$
\begin{aligned}
\dot{V}_{i-1} &= \nabla V_{i-1}^T (f + gu_i) \\
&= -\mathcal{L}(V_{i-1}, u_i) - r(x, u_i) \\
&\leq -q(x)
\end{aligned}
$$

Therefore, u_i is globally stabilizing, since V_{i-1} is a well-defined Lyapunov function for the system. In addition, we have

$$
V_i(x_0) \geq \int_0^\infty r(x, u_i) dt > 0, \forall x_0 \neq 0 \tag{4.30}
$$

Similarly as in (4.29), we can show

$$
V^*(x_0) \leq V_i(x_0) \leq V_{i-1}(x_0), \forall x_0 \in \mathbb{R}^n, \tag{4.31}
$$

and conclude that $V_i \in \mathcal{P}$.

(3) The two inequalities have been proved in (4.31).

(4) By (3), for each $x \in \mathbb{R}^n$, the sequence $\{V_i(x)\}_{i=0}^\infty$ is monotonically decreasing with 0 as its lower bound. Therefore, the limit exists, that is, there exists $V_\infty(x)$, such that $\lim_{i \to \infty} V_i(x) = V_\infty(x)$. Let $\{p_i\}_{i=1}^\infty$ be the sequence such that $V_i = p_i^T [x]_{2,2r}$. Then, we know $\lim_{i \to \infty} p_i = p_\infty \in \mathbb{R}^{n_{2r}}$, and therefore $V_\infty = p_\infty^T [x]_{2,2r}$. Also, it is easy to show $V^* \leq V_\infty \leq V_0$. Hence, $V_\infty \in \mathbb{R}[x]_{2,2r} \cap \mathcal{P}$.

(5) Let

$$
u_\infty = -\frac{1}{2} R^{-1} g^T \nabla V_\infty \tag{4.32}
$$

Then, by (4),

$$
\mathcal{H}(V_\infty) = -\mathcal{L}(V_\infty, u_\infty) \leq 0 \tag{4.33}
$$

which implies that V_∞ is a feasible solution to Problem 4.2.1. Then, the inequalities in (5) can be obtained by the fourth property in Theorem 4.2.3. ∎

The proof is thus complete.

Remark 4.3.5 *Notice that Assumption 4.3.1 holds for any controllable linear time-invariant system. For general nonlinear systems, the search of such a pair (V_0, u_1) is not trivial, because it amounts to solving some bilinear matrix inequalities (BMI) [23]. However, this problem has been actively studied in recent years, and several applicable approaches have been developed. For example, a Lyapunov-based approach utilizing state-dependent linear matrix inequalities has been studied in [23]. This method has been generalized to uncertain nonlinear polynomial systems in [36] and [10]. In [9] and [16], the authors proposed a solution for a stochastic HJB. It* *is shown that this method gives a control Lyapunov function for the deterministic system.*

Remark 4.3.6   *Notice that the control policies considered in the proposed algorithm can be extended to polynomial fractions. Indeed, instead of requiring $\mathcal{L}(V_0, u_1)$ to be SOS in Assumption 4.3.1, let us assume*

$$\alpha^2(x)\mathcal{L}(V_0, u_1) \text{ is SOS} \tag{4.34}$$

where $\alpha(x) > 0$ is an arbitrary polynomial. Then, the initial control policy u_1 can take the form of $u_1 = \alpha(x)^{-1}v(x)$, with $v(x)$ a column vector of polynomials. Then, we can relax the constraint (4.23) to the following:

$$\alpha^2(x)\mathcal{L}(V, u_i) \text{ is SOS} \tag{4.35}$$

and it is easy to see that the SOS-based policy iteration algorithm can still proceed with this new constraint.

Remark 4.3.7   *In addition to Remark 4.3.6, if $R(x)$ is not constant, by definition, there exists a symmetric matrix of polynomials $R^*(x)$, such that*

$$R(x)R^*(x) = \det(R(x))I_m \tag{4.36}$$

with $\det(R(x)) > 0$.
    As a result, the policy improvement step (4.25) gives

$$u_{i+1} = -\frac{1}{2\det(R(x))}R^*(x)g^T(x)\nabla V_i(x) \tag{4.37}$$

which is a vector of polynomial fractions.
    Now, if we select $\alpha(x)$ in (4.35) such that $\det(R(x))$ divides $\alpha(x)$, we see

$$\alpha^2(x)\mathcal{L}(V, u_i)$$

is a polynomial, and the proposed SOS-based policy iteration algorithm can still proceed with (4.23) replaced by (4.35). Following the same reasoning as in the proof of Theorem 4.3.4, it is easy to show that the solvability and the convergence properties of the proposed policy iteration algorithm will not be affected with the relaxed constraint (4.35).

4.4 GLOBAL ADP FOR UNCERTAIN POLYNOMIAL SYSTEMS

This section develops an online learning method based on the idea of ADP to implement the iterative scheme with real-time data, instead of identifying the system dynamics. For simplicity, here again we assume $R(x)$ is a constant matrix. The results can be straightforwardly extended to non-constant $R(x)$, using the same idea as discussed in Remark 4.3.7.

To begin with, consider the system

$$\dot{x} = f(x) + g(x)(u_i + e) \tag{4.38}$$

where u_i is a feedback control policy and e is a bounded time-varying function, known as the exploration noise, added for the learning purpose.

4.4.1 Forward Completeness

Similar as in Chapter 3, we will show that the system (4.38) is forward complete [1].

Definition 4.4.1 ([1])   *Consider system (4.38) with e as the input. The system is called* forward complete *if, for any initial condition $x_0 \in \mathbb{R}^n$ and every input signal e, the corresponding solution of system (4.38) is defined for all $t \geq 0$.*

Lemma 4.4.2   *Consider system (4.38). Suppose u_i is a globally stabilizing control policy and there exists $V_{i-1} \in \mathcal{P}$, such that*

$$\mathcal{L}(V_{i-1}, u_i) \geq 0 \tag{4.39}$$

Then, the system (4.38) is forward complete.

Proof: Under Assumptions 4.1.2 and 4.3.1, by Theorem 4.3.4 we know $V_{i-1} \in \mathcal{P}$. Then, by completing the squares, it follows that

$$
\begin{aligned}
\nabla V_{i-1}^T (f + gu_i + ge) &\leq -u_i^T R u_i - 2u_i^T R e \\
&= -(u_i + e)^T R(u_i + e) + e^T R e \\
&\leq e^T R e + V_{i-1}
\end{aligned}
$$

According to [1, Corollary 2.11], system (4.38) is forward complete.   ∎

By Lemma 4.4.2 and Theorem 4.3.4, we immediately have the following Proposition.

Proposition 4.4.3   *Under Assumptions 4.1.2 and 4.3.1, let u_i be a feedback control policy obtained at the ith iteration step in the proposed policy iteration algorithm (4.22)–(4.25) and e be a bounded time-varying function. Then, the closed-loop system (4.1) with $u = u_i + e$ is forward complete.*

4.4.2 Online Implementation

Under Assumption 4.3.1, in the SOS-based policy iteration, we have

$$\mathcal{L}(V_i, u_i) \in \mathbb{R}[x]_{2,2d}, \forall i > 1 \tag{4.40}$$

if the integer d satisfies

$$d \geq \frac{1}{2}\max\{\deg(f) + 2r - 1, \deg(g) + 2(2r - 1),$$
$$\deg(Q), 2(2r - 1) + 2\deg(g)\}$$

Also, u_i obtained from the proposed policy iteration algorithm satisfies $u_i \in \mathbb{R}[x]_{1,d}$.

Hence, there exists a constant matrix $K_i \in \mathbb{R}^{m \times n_d}$, with $n_d = \binom{n+d}{d} - 1$, such that $u_i = K_i[x]_{1,d}$. Also, suppose there exists a constant vector $p \in \mathbb{R}^{n_{2r}}$, with $n_{2r} = \binom{n+2r}{2r} - n - 1$, such that $V = p^T[x]_{2,2r}$. Then, along the solutions of the system (4.38), it follows that

$$\dot{V} = \nabla V^T(f + gu_i) + \nabla V^T ge$$
$$= -r(x, u_i) - \mathcal{L}(V, u_i) + \nabla V^T ge$$
$$= -r(x, u_i) - \mathcal{L}(V, u_i) + (R^{-1}g^T \nabla V)^T Re \tag{4.41}$$

Notice that the terms $\mathcal{L}(V, u_i)$ and $R^{-1}g^T \nabla V$ rely on f and g. Here, we are interested in solving them without identifying f and g.

To this end, notice that for the same pair (V, u_i) defined above, we can find a constant vector $l_p \in \mathbb{R}^{n_{2d}}$, with $n_{2d} = \binom{n+2d}{2d} - n - 1$, and a constant matrix $K_p \in \mathbb{R}^{m \times n_d}$, such that

$$\mathcal{L}(V, u_i) = l_p^T[x]_{2,2d} \tag{4.42}$$

$$-\frac{1}{2}R^{-1}g^T \nabla V = K_p[x]_{1,d} \tag{4.43}$$

Therefore, calculating $\mathcal{L}(V, u_i)$ and $R^{-1}g^T \nabla V$ amounts to finding l_p and K_p. Substituting (4.42) and (4.43) in (4.41), we have

$$\dot{V} = -r(x, u_i) - l_p^T[x]_{2,2d} - 2[x]_{1,d}^T K_p^T Re \tag{4.44}$$

Now, integrating the terms in (4.44) over the interval $[t, t + \delta t]$, we have

$$p^T\left[[x(t)]_{2,2r} - [x(t + \delta t)]_{2,2r}\right]$$
$$= \int_t^{t+\delta t}\left(r(x, u_i) + l_p^T[x]_{2,2d}\right.$$
$$\left. + 2[x]_{1,d}^T K_p^T Re\right) dt \tag{4.45}$$

Equation (4.45) implies that, l_p and K_p can be directly calculated by using real-time online data, without knowing the precise knowledge of f and g.

To see how it works, let us define the following matrices: $\sigma_e \in \mathbb{R}^{n_{2d}+mn_d}$, $\Phi_i \in \mathbb{R}^{q_i \times (n_{2d}+mn_d)}$, $\Xi_i \in \mathbb{R}^{q_i}$, $\Theta_i \in \mathbb{R}^{q_i \times n_{2r}}$, such that

$$\sigma_e = -\left[[x]_{2,2d}^T \quad 2[x]_{1,d}^T \otimes e^T R \right]^T,$$

$$\Phi_i = \left[\int_{t_{0,i}}^{t_{1,i}} \sigma_e dt \quad \int_{t_{1,i}}^{t_{2,i}} \sigma_e dt \quad \cdots \quad \int_{t_{q_i-1,i}}^{t_{q_i,i}} \sigma_e dt \right]^T,$$

$$\Xi_i = \left[\int_{t_{0,i}}^{t_{1,i}} r(x, u_i) dt \quad \int_{t_{1,i}}^{t_{2,i}} r(x, u_i) dt \quad \cdots \right.$$
$$\left. \int_{t_{q_i-1,i}}^{t_{q_i,i}} r(x, u_i) dt \right]^T,$$

$$\Theta_i = \left[[x]_{2,2r}|_{t_{0,i}}^{t_{1,i}} \quad [x]_{2,2r}|_{t_{1,i}}^{t_{2,i}} \quad \cdots \quad [x]_{2,2r}|_{t_{q_i-1,i}}^{t_{q_i,i}} \right]^T.$$

Then, (4.45) implies

$$\Phi_i \begin{bmatrix} l_p \\ \text{vec}(K_p) \end{bmatrix} = \Xi_i + \Theta_i p \tag{4.46}$$

Notice that any pair of (l_p, K_p) satisfying (4.46) will satisfy the constraint between l_p and K_p as implicitly indicated in (4.44) and (4.45).

Assumption 4.4.4 *For each $i = 1, 2, \ldots$, there exists an integer q_{i0}, such that the following rank condition holds*

$$\text{rank}(\Phi_i) = n_{2d} + mn_d, \tag{4.47}$$

if $q_i \geq q_{i0}$.

Remark 4.4.5 *This rank condition (4.47) is in the spirit of persistent excitation (PE) in adaptive control (e.g. [11, 31]) and is a necessary condition for parameter convergence.*

Given $p \in \mathbb{R}^{n_{2r}}$ and $K_i \in \mathbb{R}^{m \times n_d}$, suppose Assumption 4.4.4 is satisfied and $q_i \geq q_{i0}$ for all $i = 1, 2, \ldots$ Then, it is easy to see that the values of l_p and K_p can be uniquely determined from (4.45). Indeed,

$$\begin{bmatrix} l_p \\ \text{vec}(K_p) \end{bmatrix} = \left(\Phi_i^T \Phi_i \right)^{-1} \Phi_i^T (\Xi_i + \Theta_i p) \tag{4.48}$$

Now, the ADP-based online learning method is given below, and a flowchart is provided in Figure 4.3.

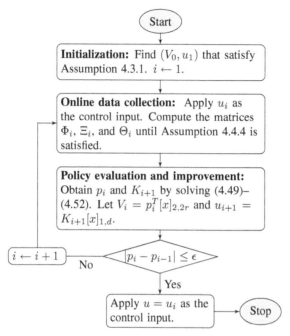

FIGURE 4.3   Flowchart of the GADP algorithm. Jiang, 2015. Reproduced with permission of IEEE.

Algorithm 4.4.6   *GADP Algorithm for polynomial systems*

(1) *Initialization:*
    Find a pair (V_0, u_1) that satisfy Assumption 4.3.1. Let p_0 be the constant vector such that $V_0 = p_0^T [x]_{2,2r}$. Let $i \leftarrow 1$.
(2) *Collect online data:*
    Apply $u = u_i + e$ to the system and compute the data matrices Φ_i, Ξ_i, and Θ_i, until Φ_i is of full-column rank.
(3) *Policy evaluation and improvement:*
    Find an optimal solution (p_i, K_{i+1}) to the following SOS program

$$\min_{p, K_p} c^T p \tag{4.49}$$

$$\text{s.t. } \Phi_i \begin{bmatrix} l_p \\ \text{vec}(K_p) \end{bmatrix} = \Xi_i + \Theta_i p \tag{4.50}$$

$$l_p^T [x]_{2,2d} \text{ is SOS} \tag{4.51}$$

$$(p_{i-1} - p)^T [x]_{2,2r} \text{ is SOS} \tag{4.52}$$

where $c = \int_\Omega [x]_{2,2r} dx$.

Then, denote $V_i = p_i^T[x]_{2,2r}$, $u_{i+1} = K_{i+1}[x]_{1,d}$, *and go to Step (2) with* $i \leftarrow i + 1$.

Theorem 4.4.7   *Under Assumptions 4.1.1, 4.3.1, and 4.4.4, the following properties hold.*

(1) *The optimization problem (4.49)–(4.52) has a nonempty feasible set.*

(2) *The sequences* $\{V_i\}_{i=1}^{\infty}$ *and* $\{u_i\}_{i=1}^{\infty}$ *satisfy the properties (2)–(5) in Theorem 4.3.4.*

Proof: Given $p_i \in \mathbb{R}^{n_{2r}}$, there exists a constant matrix $K_{p_i} \in \mathbb{R}^{m \times n_d}$ such that (p_i, K_{p_i}) is a feasible solution to the optimization problem (4.49)–(4.52), if and only if p_i is a feasible solution to the SOS program (4.22)–(4.24). Therefore, by Theorem 4.3.4, (1) holds. In addition, since the two optimization problems share the identical objective function, we know that if (p_i, K_{p_i}) is a feasible solution to the optimization problem (4.49)–(4.52), p_i is also an optimal solution to the SOS program (4.22)–(4.24). Hence, the theorem can be obtained from Theorem 4.3.4.   ∎

Remark 4.4.8   *By Assumption 4.3.1, u_1 must be a globally stabilizing control policy. Indeed, if it is only locally stabilizing, there are two reasons why the algorithm may not proceed. First, with a locally stabilizing control law, there is no way to guarantee the forward completeness of solutions, or avoid finite escape, during the learning phase. Second, the region of attraction associated with the new control policy is not guaranteed to be global. Such a counterexample is $\dot{x} = \theta x^3 + u$ with unknown positive θ, for which the choice of a locally stabilizing controller $u_1 = -x$ does not lead to a globally stabilizing suboptimal controller.*

4.5 EXTENSION FOR NONLINEAR NON-POLYNOMIAL SYSTEMS

In this section, we extend the proposed global ADP method to deal with an enlarged class of nonlinear systems. First, we will give an illustrative example to show that the SOS condition is conservative for general nonlinear functions. Second, a generalized parametrization method is proposed. Third, a less conservative sufficient condition will be derived to assure the non-negativity of a given nonlinear function. Fourth, an SDP-based implementation for the proposed policy iteration technique will be presented. Finally, an online learning method will be developed.

4.5.1 An Illustrative Example

The implementation method via SOS programs developed in Section 4.4 can efficiently handle nonlinear polynomial systems. The results could also be trivially extended to real trigonometric polynomials [3]. However, the SOS-like constraint may be too conservative to be used as a sufficient condition for non-negativity of general nonlinear functions. To see this, consider the following illustrative example:

$$f(x) = ax^2 + bx\sin x + c\sin^2 x = \begin{bmatrix} x \\ \sin x \end{bmatrix}^T P \begin{bmatrix} x \\ \sin x \end{bmatrix} \qquad (4.53)$$

Apparently, a symmetric matrix P can be uniquely determined from the constants a, b, and c. Similar to the polynomial case, we know $f(x)$ is positive semidefinite, if P is positive semidefinite. However, for the cases of $a = 1$, $b = 0$, $c = -0.5$, or $a = 0$, $b = 1$, $c = 0$, it is easy to check $f(x)$ is positive semidefinite, yet in both cases we have either $P = \begin{bmatrix} 1 & 0 \\ 0 & -0.5 \end{bmatrix}$ or $P = \begin{bmatrix} 0 & 0.5 \\ 0.5 & 0 \end{bmatrix}$, which are not positive semidefinite matrices.

This illustrative example shows that, instead of searching for a positive semidefinite matrix P, a less conservative sufficient condition for the non-negativity of more general nonlinear functions is desired. Deriving this condition and developing a global ADP method for more general nonlinear systems are the main objectives of this section.

4.5.2 Generalized Parametrization

Assumption 4.5.1   *The function* $f : \mathbb{R}^n \to \mathbb{R}^n$ *considered in system (4.1) can be parameterized as*

$$f = A\sigma \qquad (4.54)$$

where $A \in \mathbb{R}^{n \times l}$ *is an uncertain constant matrix, and* $\sigma = [\sigma_1(x), \sigma_2(x), \dots, \sigma_l(x)]^T$ *is a vector of locally Lipschitz, piecewise-continuous, and linearly independent functions, satisfying* $\sigma_i(0) = 0$, $\forall i = 1, 2, \dots, l$.

Now, we restrict each feasible solution to Problem 4.2.1 to take the form of $V(x) = \phi^T(x)P\phi(x)$, where $P \in \mathbb{R}^{N \times N}$ is a constant matrix and $\phi = [\phi_1(x), \phi_2(x), \dots, \phi_N(x)]^T$ is a vector of continuously differentiable, linearly independent, functions vanishing at the origin. To begin with, let us impose the following assumption.

Assumption 4.5.2   *The following properties are true for (4.54).*

(1) *For each* $i = 1, 2, \dots, N$; $j = 1, 2, \dots, N$; *and* $k = 1, 2, \dots, n$; *we have*

$$\frac{\partial(\phi_i\phi_j)}{\partial x_k} \in \text{span}\{\sigma_1, \sigma_2, \dots, \sigma_l\}.$$

(2) *Let g_i be the ith column of $g(x)$, with $i = 1, 2, \ldots, m$. Then,*

$$g_i^T \nabla(\phi_i \phi_j) \in \text{span}\{\sigma_1, \sigma_2, \ldots, \sigma_l\}.$$

(3) *The weighting function $q(x)$ defined in (4.2) is positive definite and satisfies*

$$q(x) \in \text{span}\{\sigma_1^2, \sigma_1 \sigma_2, \ldots, \sigma_i \sigma_j, \ldots, \sigma_l^2\}. \tag{4.55}$$

Notice that Assumption 4.5.2 is not restrictive, and can be satisfied by expanding the basis functions. Indeed, if (1) and (2) in Assumption 4.5.2 are not satisfied, we can always find locally Lipschitz functions $\sigma_{l+1}(x), \sigma_{l+2}(x), \ldots, \sigma_{l+s}(x)$, such that $\sigma_1, \sigma_2, \ldots, \sigma_{l+s}$ are linearly independent and vanishing at the origin, satisfying $\dfrac{\partial(\phi_i \phi_j)}{\partial x_k} \in \text{span}\{\sigma_1, \sigma_2, \ldots, \sigma_{l+s}\}$ and $g_i^T \nabla(\phi_i \phi_j) \in \text{span}\{\sigma_1, \sigma_2, \ldots, \sigma_{l+s}\}$. Then, the decomposition (4.54) can be rewritten as

$$f(x) = \tilde{A} \tilde{\sigma} \tag{4.56}$$

where $\tilde{A} = \begin{bmatrix} A & \mathbf{0}_{n \times s} \end{bmatrix}$ and $\tilde{\sigma} = [\sigma_1, \sigma_2, \ldots, \sigma_{l+s}]^T$.

Also, to satisfy (3) in Assumption 4.5.2, we can select $q(x)$ such that $\sqrt{q(x)}$ is locally Lipschitz and positive definite. If (4.55) does not hold, we can further define $\hat{\sigma} = [\sigma_1, \sigma_2, \ldots, \sigma_l, \sqrt{q(x)}]$. Then, clearly, all the elements in $\hat{\sigma}$ are linearly independent, and the decomposition (4.54) can be rewritten as $f = \hat{A}\hat{\sigma}$, where $\hat{A} = [A \ \mathbf{0}_{n \times 1}]$.

4.5.3 A Sufficient Condition for Non-Negativity

Define $\{\bar{\sigma}_1, \bar{\sigma}_2, \ldots, \bar{\sigma}_{l_1}\}$ as the largest linearly independent subset of $\{\sigma_1^2, \sigma_1 \sigma_2, \ldots, \sigma_i \sigma_j, \ldots, \sigma_l^2\}$, and $\{\bar{\phi}_1, \bar{\phi}_2, \ldots, \bar{\phi}_{N_1}\}$ as the largest linearly independent subset of $\{\phi_1^2, \phi_1 \phi_2, \ldots, \phi_i \phi_j, \ldots, \phi_N^2\}$.

Then, if $W \in \text{span}\{\phi_1^2, \phi_1 \phi_2, \ldots, \phi_N^2\}$ and $\delta \in \text{span}\{\phi_1^2, \sigma_1 \sigma_2, \ldots, \sigma_l^2\}$, there exist uniquely constant vectors $p \in \mathbb{R}^{N_1}$ and $h \in \mathbb{R}^{l_1}$, such that $W = p^T \bar{\phi}$ and $\delta = h^T \bar{\sigma}$, where $\bar{\phi} = [\bar{\phi}_1, \bar{\phi}_2, \ldots, \bar{\phi}_{N_1}]^T$ and $\bar{\sigma} = [\bar{\sigma}_1, \bar{\sigma}_2, \ldots, \bar{\sigma}_{l_1}]^T$.

Using the above-mentioned parametrization method, we now show that it is possible to decide if W and δ are positive semidefinite functions, by studying the coefficient vectors p and h.

Without loss of generality, we assume the following properties of $\bar{\phi}_i$:

(1) For $i = 1, 2, \ldots, N_2$, we have $\bar{\phi}_i \geq 0$, with N_2 an integer satisfying $1 \leq N_2 \leq N_1$.

(2) There exist integers i_r and j_r with $r = 1, 2, \ldots, N_3$, such that $1 \leq i_r, j_r \leq N_2$, $i_r \neq j_r$ and $\bar{\phi}_{i_r} \geq \bar{\phi}_{j_r}$.

Definition 4.5.3   *For any $p \in \mathbb{R}^{N_1}$, we say $p \in \mathbb{S}_\phi^+$ if and only if there exist constants $\gamma_1, \gamma_2, \dots, \gamma_{N_2} \geq 0$, $\alpha_1, \alpha_2, \dots, \alpha_{N_3} \geq 0$, $\beta_1, \beta_2, \dots, \beta_{N_3}$, and a symmetric positive semidefinite matrix $P \in \mathbb{R}^{N \times N}$, such that $\alpha_i + \beta_i \geq 0$, for $i = 1, 2, \dots, N_3$, and*

$$p = M_\phi^T \mathrm{vec}(P) + \begin{bmatrix} \gamma_1 \\ \gamma_2 \\ \vdots \\ \gamma_{N_2} \\ \mathbf{0}_{N_1 - N_2} \end{bmatrix} + \sum_{r=1}^{N_3} \left(\begin{bmatrix} \mathbf{0}_{i_r - 1} \\ \alpha_r \\ \mathbf{0}_{N_1 - i_r} \end{bmatrix} + \begin{bmatrix} \mathbf{0}_{j_r - 1} \\ \beta_r \\ \mathbf{0}_{N_1 - j_r} \end{bmatrix} \right) \tag{4.57}$$

where $M_\phi \in \mathbb{R}^{N^2 \times N_1}$ is a constant matrix satisfying $M_\phi \bar{\phi} = \phi \otimes \phi$.
In addition, we say W belongs to the set $\mathbb{S}_\phi^+[x]$ if and only if there exists $p \in \mathbb{S}_\phi^+$, such that $W = p^T \bar{\phi}$.

Lemma 4.5.4   *If $p \in \mathbb{S}_\phi^+$, then $p^T \bar{\phi}$ is positive semidefinite.*

Proof: By definition, if $p \in \mathbb{S}_\phi^+$, it follows that

$$p^T \bar{\phi} = \phi^T P \phi + \sum_{i=1}^{N_2} \gamma_i \bar{\phi}_i + \sum_{r=1}^{N_3} (\alpha_r \bar{\phi}_{i_r} + \beta_r \bar{\phi}_{j_r})$$

$$\geq \sum_{r=1}^{N_3} (\alpha_r \bar{\phi}_{i_r} - |\beta_r| \bar{\phi}_{j_r}) \geq \sum_{r=1}^{N_3} (\alpha_r - |\beta_r|) \bar{\phi}_{j_r}$$

$$\geq 0$$

The proof is complete.   ■

In the same way, we can find two sets $\mathbb{S}_\sigma^+$ and $\mathbb{S}_\sigma^+[x]$, such that the following implications hold

$$h \in \mathbb{S}_\sigma^+ \Leftrightarrow h^T \bar{\phi} \in \mathbb{S}_\sigma^+[x] \Rightarrow h^T \bar{\phi} \geq 0 \tag{4.58}$$

4.5.4 Generalized Policy Iteration

To begin with, let us assume the existence of a parameterized admissible control policy.

Assumption 4.5.5   *There exist $p_0 \in \mathbb{S}_\phi^+$ and $K_1 \in \mathbb{R}^{m \times l_1}$, such that $V_0 = p_0^T \bar{\phi}$, $u_1 = K_1 \sigma$, and $\mathcal{L}(V_0, u_1) \in \mathbb{S}_\sigma^+[x]$.*

Remark 4.5.6   *Under Assumptions 4.5.1, 4.5.2, and 4.5.5, Assumption 4.1.1 is satisfied.*

Now, let us show how the proposed policy iteration can be practically implemented. First of all, given $p \in \mathbb{R}^{N_1}$, since $u_i = K_i \sigma$, we can always find two linear mappings $\bar{\iota} : \mathbb{R}^{N_1} \times \mathbb{R}^{ml} \to \mathbb{R}^{l_1}$ and $\bar{\kappa} : \mathbb{R}^{N_1} \to \mathbb{R}^{l_1 \times ml}$, such that

$$\bar{\iota}(p, K)^T \bar{\sigma} = \mathcal{L}(p^T \bar{\phi}, K_i \sigma) \tag{4.59}$$

$$\bar{\kappa}(p)^T \bar{\sigma} = -\frac{1}{2} R^{-1} g^T \nabla(p^T \bar{\phi}) \tag{4.60}$$

Then, under Assumptions 4.1.2, 4.5.1, 4.5.2, and 4.5.5, the proposed policy iteration can be implemented as follows.

Algorithm 4.5.7   *Generalized SOS-based policy iteration algorithm*

(1) *Initialization:*
   Find $p_0 \in \mathbb{R}^{N_1}$ and $K_1 \in \mathbb{R}^{m \times l_1}$ satisfying Assumption 4.5.5, and let $i = 1$.
(2) *Policy evaluation and improvement:*
   Solve for an optimal solution (p_i, K_{i+1}) of the following problem.

$$\min_{p, K} \quad c^T p \tag{4.61}$$

$$\text{s.t.} \ \bar{\iota}(p, K_i) \in \mathbb{S}_\sigma^+ \tag{4.62}$$

$$p_{i-1} - p \in \mathbb{S}_\phi^+ \tag{4.63}$$

$$K = \bar{\kappa}(p) \tag{4.64}$$

   where $c = \int_\Omega \bar{\phi}(x) dx$. Then, denote $V_i = p_i^T \bar{\phi}$ and $u_{i+1} = K_{i+1} \sigma$.
(3) *Go to Step (2) with i replaced by $i + 1$.*

Some useful facts about the above-mentioned policy iteration algorithm are summarized in the following theorem, of which the proof is omitted, since it is nearly identical to the proof of Theorem 4.3.4.

Theorem 4.5.8   *Under Assumptions 4.1.2, 4.5.1, 4.5.2, and 4.5.5, the following are true, for $i = 1, 2, \ldots$*

(1) *The optimization problem (4.61)–(4.64) has a nonempty feasible set.*
(2) *The closed-loop system comprised of (4.1) and $u = u_i(x)$ is globally asymptotically stable at the origin.*
(3) *$V_i \in \mathcal{P}$. In addition, $V^*(x_0) \le V_i(x_0) \le V_{i-1}(x_0), \forall x_0 \in \mathbb{R}^n$.*
(4) *There exists $p_\infty \in \mathbb{R}^{N_1}$, such that $\lim_{i \to \infty} V_i(x_0) = p_\infty^T \bar{\phi}(x_0), \forall x_0 \in \mathbb{R}^n$.*

(5) *Along the solutions of the system composed of (4.1) and (4.7), it follows that*

$$0 \le p_\infty^T \bar{\phi}(x_0) - V^*(x_0) \le - \int_0^\infty \mathcal{H}(p_\infty^T \bar{\phi}(x(t)))dt. \tag{4.65}$$

4.5.5 Online Implementation via Global ADP

Let $V = p^T \bar{\phi}$. Similar as in Section 4.4, over the interval $[t, t + \delta t]$, we have

$$p^T \left[\bar{\phi}(x(t)) - \bar{\phi}(x(t + \delta t)) \right]$$
$$= \int_t^{t+\delta t} \left[r(x, u_i) + \bar{\iota}(p, K_i)^T \bar{\sigma} + 2\sigma^T \bar{\kappa}(p)^T Re \right] dt \tag{4.66}$$

Therefore, (4.66) shows that, given $p \in \mathbb{R}^{N_1}$, $\bar{\iota}(p, K_i)$ and $\bar{\kappa}(p)$ can be directly obtained by using real-time measurements, without knowing the precise knowledge of f and g.

Indeed, define

$$\bar{\sigma}_e = - \left[\bar{\sigma}^T \quad 2\sigma^T \otimes e^T R \right]^T \in \mathbb{R}^{l_1 + ml},$$

$$\bar{\Phi}_i = \left[\int_{t_{0,i}}^{t_{1,i}} \bar{\sigma}_e dt \quad \int_{t_{1,i}}^{t_{2,i}} \bar{\sigma}_e dt \quad \cdots \quad \int_{t_{q_i-1,i}}^{t_{q_i,i}} \bar{\sigma}_e dt \right]^T \in \mathbb{R}^{q_i \times (l_1 + ml)},$$

$$\bar{\Xi}_i = \left[\int_{t_{0,i}}^{t_{1,i}} r(x, u_i) dt \quad \int_{t_{1,i}}^{t_{2,i}} r(x, u_i) dt \quad \cdots \quad \int_{t_{q_i-1,i}}^{t_{q_i,i}} r(x, u_i) dt \right]^T \in \mathbb{R}^{q_i},$$

$$\bar{\Theta}_i = \left[\bar{\phi}(x)|_{t_{0,i}}^{t_{1,i}} \quad \bar{\phi}(x)|_{t_{1,i}}^{t_{2,i}} \quad \cdots \quad \bar{\phi}(x)|_{t_{q_i-1,i}}^{t_{q_i,i}} \right]^T \in \mathbb{R}^{q_i \times N_1}.$$

Then, (4.66) implies

$$\bar{\Phi}_i \left[\begin{array}{c} \bar{\iota}(p, K_i) \\ \mathrm{vec}(\bar{\kappa}(p)) \end{array} \right] = \bar{\Xi}_i + \bar{\Theta}_i p \tag{4.67}$$

Assumption 4.5.9   *For each $i = 1, 2, \ldots$, there exists an integer q_{i0}, such that, when $q_i \ge q_{i0}$, the following rank condition holds.*

$$\mathrm{rank}(\bar{\Phi}_i) = l_1 + ml \tag{4.68}$$

Let $p \in \mathbb{R}^{N_1}$ and $K_i \in \mathbb{R}^{m \times l}$. Suppose Assumption 4.5.9 holds and assume $q_i \ge q_{i0}$, for $i = 1, 2, \ldots$ Then, $\bar{\iota}(p, K_i) = h$ and $\bar{\kappa}(p) = K$ can be uniquely determined by

$$\left[\begin{array}{c} h \\ \mathrm{vec}(K) \end{array} \right] = (\bar{\Phi}_i^T \bar{\Phi}_i)^{-1} \bar{\Phi}_i^T (\bar{\Xi}_i + \bar{\Theta}_i p) \tag{4.69}$$

Now, we are ready to develop the ADP-based online implementation algorithm for the proposed policy iteration method.

Algorithm 4.5.10   *GADP algorithm for non-polynomial systems*

(1) *Initialization:*
    Let p_0 and K_1 satisfy Assumption 4.5.5, and let $i = 1$.
(2) *Collect online data:*
    Apply $u = u_i + e$ to the system and compute the data matrices $\bar{\Phi}_i$, $\bar{\Xi}_i$, and $\bar{\Theta}_i$, until the rank condition (4.68) is satisfied.
(3) *Policy evaluation and improvement:*
    Find an optimal solution (p_i, h_i, K_{i+1}) to the following optimization problem

$$\min_{p,h,K} c^T p \tag{4.70}$$

$$\text{s.t.} \begin{bmatrix} h \\ \text{vec}(K) \end{bmatrix} = \left(\bar{\Phi}_i^T \bar{\Phi}_i \right)^{-1} \bar{\Phi}_i^T \left(\bar{\Xi}_i + \bar{\Theta}_i p \right) \tag{4.71}$$

$$h \in \mathbb{S}_\sigma^+ \tag{4.72}$$

$$p_{i-1} - p \in \mathbb{S}_\phi^+ \tag{4.73}$$

    Then, denote $V_i = p_i \bar{\phi}$ and $u_{i+1} = K_{i+1} \bar{\sigma}$
(4) *Go to Step (2) with $i \leftarrow i + 1$.*

Properties of the above algorithm are summarized in the following corollary.

Corollary 4.5.11   *Under Assumptions 4.1.2, 4.5.1, 4.5.2, 4.5.5, and 4.5.9, the algorithm enjoys the following properties.*

(1) *The optimization problem (4.70)–(4.73) has a feasible solution.*
(2) *The sequences $\{V_i\}_{i=1}^\infty$ and $\{u_i\}_{i=1}^\infty$ satisfy the properties (2)–(5) in Theorem 4.5.8.*

4.6 APPLICATIONS

This section covers the numerical simulation for four different examples. First, we apply the SOS-based policy iteration to the optimal control design of the car suspension system studied in Chapter 3. We will show that, with the full knowledge of the system dynamics, a suboptimal and globally stabilizing control policy can be obtained. Second, we apply the proposed GADP method to an illustrative scalar system and a fault-tolerant control problem [36], both of the systems are uncertain polynomial systems. Finally, we use the non-polynomial extension described in Section 4.5 to achieve the online optimal control of an inverted pendulum.

4.6.1 Car Suspension System

Recall from Chapter 3 the mathematical model of a car suspension system:

$$\dot{x}_1 = x_2 \tag{4.74}$$

$$\dot{x}_2 = -\frac{k_s(x_1-x_3) + k_n(x_1-x_3)^3 + b_s(x_2-x_4) - cu}{m_b} \tag{4.75}$$

$$\dot{x}_3 = x_3 \tag{4.76}$$

$$\dot{x}_4 = \frac{k_s(x_1-x_3) + k_n(x_1-x_3)^3 + b_s(x_2-x_4) - k_t x_3 - cu}{m_w} \tag{4.77}$$

where x_1, x_2, and m_b denote respectively the position, velocity, and mass of the car body; x_3, x_4, and m_w represent respectively the position, velocity, and mass of the wheel assembly; k_t, k_s, k_n, and b_s are the tyre stiffness, the linear suspension stiffness, the nonlinear suspension stiffness, and the damping rate of the suspension; c is a constant relating the control signal to input force.

Again, we assume that $m_b \in [250, 350]$, $m_w \in [55, 65]$, $b_s \in [900, 1100]$, $k_s \in [15000, 17000]$, $k_n = k_s/10$, $k_t \in [180000, 200000]$, and $c > 0$. Hence, the uncontrolled system is globally asymptotically stable at the origin. The same performance index is considered here, and it will be reduced to achieve better performance of the closed-loop system.

$$J(x_0; u) = \int_0^\infty \left(\sum_{i=1}^4 x_i^2 + u^2 \right) dt \tag{4.78}$$

In addition to global asymptotic stability, we are also interested in improving the system performance when the initial condition is in the following set.

$$\Omega = \{x | x \in \mathbb{R}^4 \text{ and } |x_1| \le 0.5, |x_2| \le 10, |x_3| \le 0.5, |x_4| \le 10\}$$

We applied the SOS-based policy iteration and the approximation control policy is obtained after 9 iterations. To show the improvement of the performance, we simulated an impulse disturbance in the beginning of the simulation. Then, we compared state trajectories of the closed-loop system with the uncontrolled system trajectories, as shown in Figure 4.4. The initial value function and the improved one are compared in Figure 4.5. Notice that, compared with the semi-globally stabilizing control policy in Chapter 3, the control policy obtained here globally asymptotically stabilizes the car suspension system.

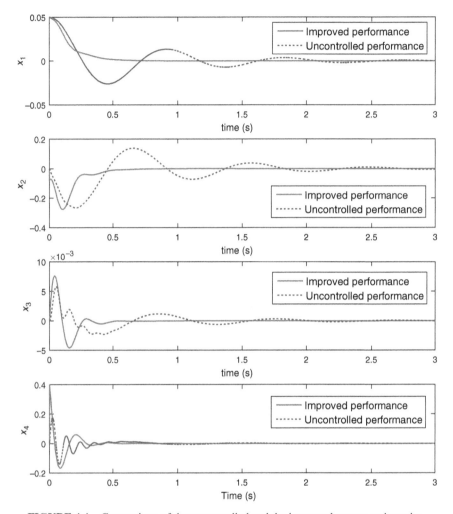

FIGURE 4.4   Comparison of the uncontrolled and the improved system trajectories.

4.6.2   A Scalar Nonlinear Polynomial System

Consider the following polynomial system

$$\dot{x} = ax^2 + bu \tag{4.79}$$

where $x \in \mathbb{R}$ is the system state, $u \in \mathbb{R}$ is the control input, and a and b, satisfying $a \in [0, 0.05]$ and $b \in [0.5, 1]$, are uncertain constants. The cost to be minimized is defined as

$$J(x_0, u) = \int_0^\infty (0.01x^2 + 0.01x^4 + u^2)dt \tag{4.80}$$

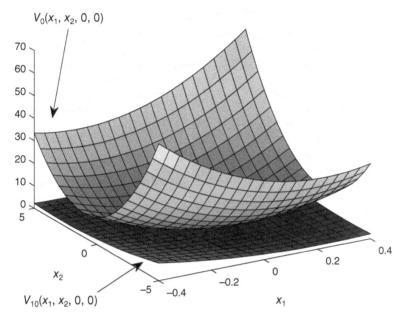

$V_0(x_1, x_2, 0, 0)$

$V_{10}(x_1, x_2, 0, 0)$

FIGURE 4.5 Comparison of the value functions before and after learning. Jiang, 2015. Reproduced with permission of IEEE.

An initial stabilizing control policy can be selected as $u_1 = -0.1x - 0.1x^3$, which globally asymptotically stabilizes system (4.79), for any a and b satisfying the given range. Further, it is easy to see that $V_0 = 10(x^2 + x^4)$ and u_1 satisfy Assumption 4.3.1 with $r = 2$. In addition, we set $d = 3$.

Only for the purpose of simulation, we set $a = 0.01$, $b = 1$, and $x(0) = 2$. Ω is specified as $\Omega = \{x | x \in \mathbb{R} \text{ and } |x| \leq 1\}$. The proposed global ADP method is applied with the control policy updated after every 5 seconds, and convergence is attained after 5 iterations, when $|p_i - p_{i-1}| \leq 10^{-3}$. Hence, the constant vector c in the objective function (4.49) is computed as $c = [\frac{2}{3} \ 0 \ \frac{2}{5}]^T$. The exploration noise is set to be $e = 0.01(\sin(10t) + \sin(3t) + \sin(100t))$, which is turned off after the fourth iteration.

The simulated state trajectory is shown in Figure 4.6, where the control policy is updated every 5 s until convergence is attained. The suboptimal control policy and the cost function obtained after 4 iterations are $V^* = 0.1020x^2 + 0.007x^3 + 0.0210x^4$ and $u^* = -0.2039x - 0.02x^2 - 0.0829x^3$.

For comparison purpose, the exact optimal cost and the control policy are given below.

$$V^* = \frac{x^3}{150} + \frac{(\sqrt{101x^2 + 100})^3}{15150} - \frac{20}{303}$$

$$u^* = -\frac{x^2\sqrt{101x^2 + 100} + 101x^4 + 100x^2}{100\sqrt{101x^2 + 100}}$$

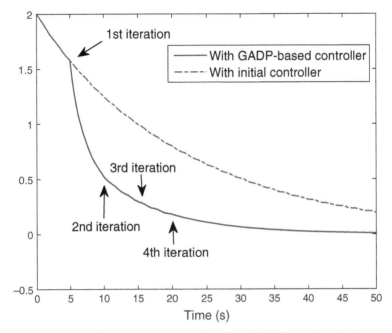

FIGURE 4.6   Simulated system state trajectories. Jiang, 2015. Reproduced with permission of IEEE.

Figures 4.7 and 4.8 show the comparison of the suboptimal control policy with respect to the exact optimal control policy and the initial control policy. The actual control signal during the simulation is shown in Figure 4.9.

4.6.3 Fault-Tolerant Control

Consider the following example [36]

$$
\begin{bmatrix} \dot{x}_1 \\ \dot{x}_2 \end{bmatrix} = \begin{bmatrix} -x_1^3 - x_1 x_2^2 + x_1 x_2 \\ x_1 + 2x_2 \end{bmatrix} + \begin{bmatrix} 0 & \beta_1 \\ \beta_2 & \beta_1 \end{bmatrix} u \tag{4.81}
$$

where $\beta_1, \beta_2 \in [0.5, 1]$ are the uncertain parameters. This system can be considered as having a loss-of-effectiveness fault [36], when the actuator gains are smaller than the commanded position ($\beta_1 = \beta_2 = 1$).

Using SOS-related techniques, it has been shown in [36] that the following robust control policy can globally asymptotically stabilize the system (4.81) at the origin.

$$
u_1 = \begin{bmatrix} u_{1,1} \\ u_{1,2} \end{bmatrix} = \begin{bmatrix} 10.283x_1 - 13.769x_2 \\ -10.7x_1 - 3.805x_2 \end{bmatrix} \tag{4.82}
$$

Here, we take into account optimality of the closed-loop system.

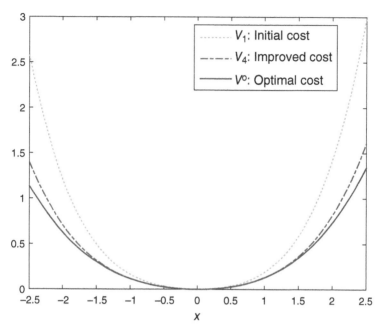

FIGURE 4.7 Comparison of the value functions. Jiang, 2015. Reproduced with permission of IEEE.

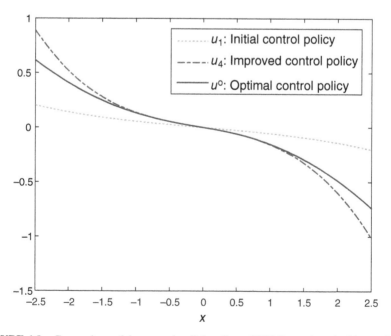

FIGURE 4.8 Comparison of the control policies. Jiang, 2015. Reproduced with permission of IEEE.

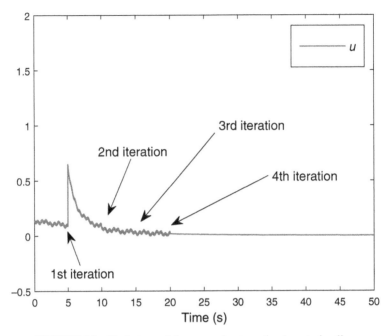

FIGURE 4.9   Trajectory of the approximate optimal control policy.

Our control objective is to improve the control policy using the proposed ADP method such that we can reduce the following cost

$$J(x_0, u) = \int_0^\infty (x_1^2 + x_2^2 + u^T u)dt \qquad (4.83)$$

and at the same time guarantee global asymptotic stability for the closed-loop system. In particular, we are interested to improve the performance of the closed-loop system in the set $\Omega = \{x | x \in \mathbb{R}^2 \text{ and } |x| \le 1\}$.

By solving the following feasibility problem using SOSTOOLS [20, 22]

$$V \in \mathbb{R}[x]_{2,4} \qquad (4.84)$$
$$\mathcal{L}(V, u_1) \text{ is SOS}, \quad \forall \beta_1, \beta_2 \in [0.5, 1] \qquad (4.85)$$

we have obtained a polynomial function as follows

$$V_0 = 17.6626x_1^2 - 18.2644x_1x_2 + 16.4498x_2^2$$
$$-0.1542x_1^3 + 1.7303x_1^2x_2 - 1.0845x_1x_2^2$$
$$+0.1267x_2^3 + 3.4848x_1^4 - 0.8361x_1^3x_2$$
$$+4.8967x_1^2x_2^2 + 2.3539x_2^4$$

which, together with (4.82), satisfies Assumption 4.3.1.

Only for simulation purpose, we set $\beta_1 = 0.7$ and $\beta_2 = 0.6$ and set the initial conditions as $x_1(0) = 1$ and $x_2(0) = -2$. The proposed online learning scheme is applied to update the control policy every 4 s for seven times. The exploration noise is the sum of sinusoidal waves with different frequencies, and it is turned off after the last iteration. The suboptimal and globally stabilizing control policy is $u_8 = [u_{8,1}, u_{8,2}]^T$ with

$$
\begin{aligned}
u_{8,1} = &-0.0004x_1^3 - 0.0067x_1^2 x_2 - 0.0747x_1^2 \\
&+0.0111x_1 x_2^2 + 0.0469x_1 x_2 - 0.2613x_1 \\
&-0.0377x_2^3 - 0.0575x_2^2 - 2.698x_2 \\
u_{8,2} = &-0.0005x_1^3 - 0.009x_1^2 x_2 - 0.101x_1^2 \\
&+0.0052x_1 x_2^2 - 0.1197x_1 x_2 - 1.346x_1 \\
&-0.0396x_2^3 - 0.0397x_2^2 - 3.452x_2
\end{aligned}
$$

The associated cost function is as follows:

$$
\begin{aligned}
V_8 = &\, 1.4878x_1^2 + 0.8709x_1 x_2 + 4.4963x_2^2 \\
&+0.0131x_1^3 + 0.2491x_1^2 x_2 - 0.0782x_1 x_2^2 \\
&+0.0639x_2^3 + 0.0012x_1^3 x_2 + 0.0111x_1^2 x_2^2 \\
&-0.0123x_1 x_2^3 + 0.0314x_2^4
\end{aligned}
$$

In Figure 4.10, we show the system state trajectories during the learning phase and the post-learning phase. At $t = 30$s, we inject an impulse disturbance through the input channel to deviate the state from the origin. Then, we compare the system response under the proposed control policy and the initial control policy given in [36]. The suboptimal value function and the original value function are compared in Figure 4.11. The improved control policy and the initial control policy are compared in Figure 4.12.

4.6.4 Inverted Pendulum: A Non-Polynomial System

Consider the following differential equations which are used to model an inverted pendulum:

$$
\dot{x}_1 = x_2 \tag{4.86}
$$

$$
\dot{x}_2 = -\frac{kl}{m}x_2 + g\sin(x_1) + \frac{1}{m}u \tag{4.87}
$$

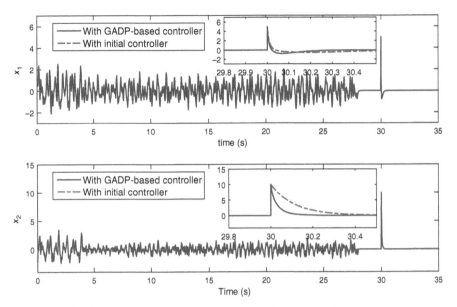

FIGURE 4.10   System state trajectories. Jiang, 2015. Reproduced with permission of IEEE.

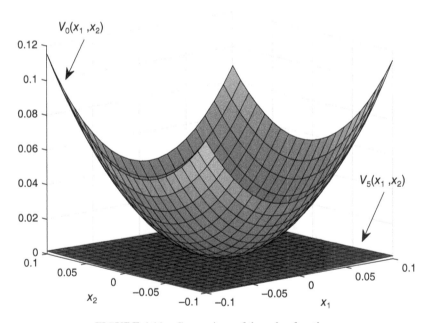

FIGURE 4.11   Comparison of the value functions.

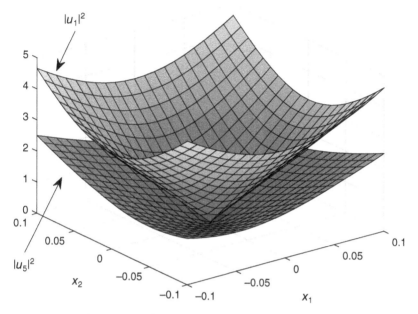

FIGURE 4.12 Comparison of the control policies.

where x_1 is the angular position of the pendulum, x_2 is the angular velocity, u is the control input, g is the gravity constant, l is the length of the pendulum, k is the coefficient of friction, and m is the mass. The design objective is to find a suboptimal and globally stabilizing control policy that can drive the state to the origin. Assume the parameters are not precisely known, but they satisfy $0.5 \leq k \leq 1.5, 0.5 \leq m \leq 1.5$, $0.8 \leq l \leq 1.2$, and $9 \leq g \leq 10$.

Notice that we can select $\phi = [x_1, x_2]^T$ and $\sigma = [x_1, x_2, \sin x_1]^T$. The cost is selected as $J(x_0, u) = \int_0^\infty (10x_1^2 + 10x^2 + u^2)dt$.

Further, set $\bar{\phi} = [x_1^2, \; x_1 x_2, \; x_2^2]^T$ and $\bar{\sigma} = [x_1^2, \; x_2^2, \; x_1 \sin x_1, \; \sin^2 x_1, \; x_2 \sin x_1,$ $x_1 x_2]^T$. Then, based on the range of the system parameters, a pair (V_0, u_1) satisfying Assumption 4.5.5 can be obtained as $u_1 = -10x_1^2 - x_2 - 15 \sin x_1$, and $V_0 = 320.1297x_1^2 + 46.3648x_1 x_2 + 22.6132x_2^2$. The coefficient vector c is defined as $c = \bar{\phi}(1, -1) + \bar{\phi}(1, 1)$.

The initial condition for the system is set to be $x_1(0) = -1.5$ and $x_2(0) = 1$. The control policy is updated after 0.5 s, until convergence is attained after 4 iterations. The exploration noise we use is the sum of sinusoidal waves with different frequencies, and it is terminated once the convergence is attained.

The resultant control policy and the cost function are $u_\infty = -20.9844x_1 - 7.5807x_2$ and $V_\infty = 86.0463x_1^2 + 41.9688x_1 x_2 + 7.5807x_2^2$. Simulation results are provided in Figure 4.13. The improvement of the value function can be seen in Figure 4.14.

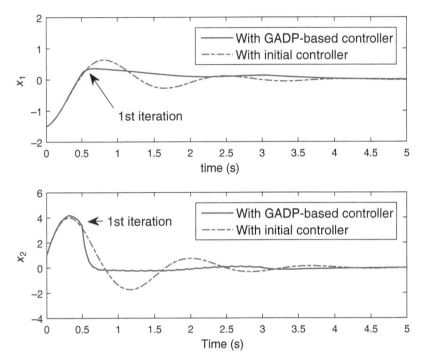

FIGURE 4.13 System state trajectories.

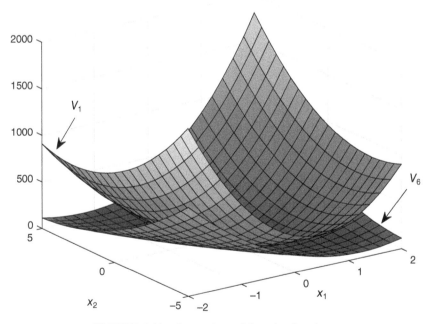

FIGURE 4.14 Comparison of the value functions.

4.7 NOTES

A common approach in ADP is to approximate the value function using a set of basis functions. Since approximation error is inevitable, the approximate optimal control policy is generally suboptimal and may even become unstable. Hence, it is of great importance to perform analysis on the approximation error and to find a bound of the value function of the closed-loop system, to which the approximate control policy is applied. This type of analysis can be conducted by studying the related HJ inequality, or the relaxed HJB mentioned in this chapter.

The relaxation method was first introduced in the approximate dynamic programming for Markov decision processes (MDPs) with finite state space [5]. A relaxed value iteration approach for MDPs is proposed in [18], where complexity is reduced by relaxing the demand for optimality. In [34] and [35], the relaxed Bellman equation of linear stochastic systems is studied and SDP is used to produce a bound on the optimal objective and a suboptimal control policy. Linear programming and polynomial approximations for MDPs are found in [27]. In [30], the Bellman equation is relaxed to an inequality and then reformulated as an SOS program. Hamilton-Jacobi (HJ) inequalities are also considered in nonlinear $\mathcal{H}_\infty$ control [8, 25, 32]. However, solving the relaxed HJB is non-trivial in general, even for polynomial systems (see, e.g., [4, 6, 19, 29, 36]). Indeed, for any polynomial with degree no less than four, deciding its non-negativity is an NP-hard (nondeterministic polynomial-time hard) problem [21]. Thanks to the developments in SOS programs [3, 21], the computational burden can be significantly reduced, if inequality constraints can be restricted to SOS constraints. In [26], a hierarchy of linear matrix inequality (LMI) relations is derived based on the theory of moments for discrete-time nonlinear stochastic polynomial systems, and the dual linear programming with the dual hierarchy of dual SDPs is used to derive control laws. In [13], conventional policy iteration for nonlinear polynomial systems is relaxed by using SOS programs, while global stability is guaranteed. In addition, the exact knowledge of the system dynamics is not required. This chapter extends the methodology proposed in [13] for a class of non-polynomial systems. Further extension may be achieved by incorporating results on semi-algebraic functions [15].

In addition to the numerical examples studied in this chapter, GADP has found application in cooperative adaptive cruise control for connected platooning vehicles [7].

REFERENCES

[1] D. Angeli and E. D. Sontag. Forward completeness, unboundedness observability, and their Lyapunov characterizations. *Systems & Control Letters*, 38(4):209–217, 1999.

[2] R. W. Beard, G. N. Saridis, and J. T. Wen. Galerkin approximations of the generalized Hamilton-Jacobi-Bellman equation. *Automatica*, 33(12):2159–2177, 1997.

[3] G. Blekherman, P. A. Parrilo, and R. R. Thomas (editors). *Semidefinite Optimization and Convex Algebraic Geometry*. SIAM, Philadelphia, PA, 2013.

[4] Z. Chen and J. Huang. Global robust stabilization of cascaded polynomial systems. *Systems & Control Letters*, 47(5):445–453, 2002.

[5] D. P. de Farias and B. Van Roy. The linear programming approach to approximate dynamic programming. *Operations Research*, 51(6):850–865, 2003.

[6] G. Franze, D. Famularo, and A. Casavola. Constrained nonlinear polynomial time-delay systems: A sum-of-squares approach to estimate the domain of attraction. *IEEE Transactions on Automatic Control*, 57(10):2673–2679, 2012.

[7] W. Gao and Z. P. Jiang. Nonlinear and adaptive suboptimal control of connected vehicles: a global adaptive dynamic programming approach. *Journal of Intelligent & Robotic Systems*, 2016. doi:10.1007/s10846-016-0395-3

[8] J. W. Helton and M. R. James. *Extending H_∞ Control to Nonlinear Systems: Control of Nonlinear Systems to Achieve Performance Objectives*. SIAM, 1999.

[9] M. B. Horowitz and J. W. Burdick. Semidefinite relaxations for stochastic optimal control policies. In: Proceedings of the 2014 American Control Conference, pp. 3006–3012, Portland, Oregon, June 1994.

[10] W.-C. Huang, H.-F. Sun, and J.-P. Zeng. Robust control synthesis of polynomial nonlinear systems using sum of squares technique. *Acta Automatica Sinica*, 39(6):799–805, 2013.

[11] P. A. Ioannou and J. Sun. *Robust Adaptive Control*. Prentice-Hall, Upper Saddle River, NJ, 1996.

[12] Y. Jiang and Z. P. Jiang. Robust adaptive dynamic programming and feedback stabilization of nonlinear systems. *IEEE Transactions on Neural Networks and Learning Systems*, 25(5):882–893, 2014.

[13] Y. Jiang and Z. P. Jiang. Global adaptive dynamic programming for continuous-time nonlinear systems. *IEEE Transactions on Automatic Control*, 60(11):2917–2929, November 2015.

[14] H. K. Khalil. *Nonlinear Systems*, 3rd ed. Prentice Hall, Upper Saddle River, NJ, 2002.

[15] J. B. Lasserre and M. Putinar. Positivity and optimization for semi-algebraic functions. *SIAM Journal on Optimization*, 20(6):3364–3383, 2010.

[16] Y. P. Leong, M. B. Horowitz, and J. W. Burdick. Optimal controller synthesis for nonlinear dynamical systems. *arXiv preprint*, arXiv:1410.0405, 2014.

[17] F. L. Lewis, D. Vrabie, and V. L. Syrmos. *Optimal Control*, 3rd ed. John Wiley & Sons, New York, 2012.

[18] B. Lincoln and A. Rantzer. Relaxing dynamic programming. *IEEE Transactions on Automatic Control*, 51(8):1249–1260, 2006.

[19] E. Moulay and W. Perruquetti. Stabilization of nonaffine systems: A constructive method for polynomial systems. *IEEE Transactions on Automatic Control*, 50(4):520–526, 2005.

[20] A. Papachristodoulou, J. Anderson, G. Valmorbida, S. Prajna, P. Seiler, and P. A. Parrilo. SOSTOOLS: Sum of squares optimization toolbox for MATLAB V3.00, 2013. Available at http://www.cds.caltech.edu/sostools

[21] P. A. Parrilo. Structured semidefinite programs and semialgebraic geometry methods in robustness and optimization. PhD Thesis, California Institute of Technology, Pasadena, California, 2000.

[22] S. Prajna, A. Papachristodoulou, and P. A. Parrilo. Introducing SOSTOOLS: A general purpose sum of squares programming solver. In: Proceedings of the 41st IEEE Conference on Decision and Control, pp. 741–746, Las Vegas, NV, 2002.

[23] S. Prajna, A. Papachristodoulou, and F. Wu. Nonlinear control synthesis by sum of squares optimization: A Lyapunov-based approach. In: Proceedings of the Asian Control Conference, pp. 157–165, Melbourne, Australia, 2004.

[24] G. N. Saridis and C.-S. G. Lee. An approximation theory of optimal control for trainable manipulators. *IEEE Transactions on Systems, Man and Cybernetics*, 9(3):152–159, 1979.

[25] M. Sassano and A. Astolfi. Dynamic approximate solutions of the HJ inequality and of the HJB equation for input-affine nonlinear systems. *IEEE Transactions on Automatic Control*, 57(10):2490–2503, October 2012.

[26] C. Savorgnan, J. B. Lasserre, and M. Diehl. Discrete-time stochastic optimal control via occupation measures and moment relaxations. In: Proceedings of the Joint 48th IEEE Conference on Decision and Control and the 28th Chinese Control Conference, pp. 519–524, Shanghai, P. R. China, 2009.

[27] P. J. Schweitzer and A. Seidmann. Generalized polynomial approximations in Markovian decision processes. *Journal of Mathematical Analysis and Applications*, 110(2):568–582, 1985.

[28] R. Sepulchre, M. Jankovic, and P. Kokotovic. *Constructive Nonlinear Control*. Springer-Verlag, New York, 1997.

[29] E. D. Sontag. On the observability of polynomial systems, I: Finite-time problems. *SIAM Journal on Control and Optimization*, 17(1):139–151, 1979.

[30] T. H. Summers, K. Kunz, N. Kariotoglou, M. Kamgarpour, S. Summers, and J. Lygeros. Approximate dynamic programming via sum of squares programming. In: Proceedings of 2013 European Control Conference (ECC), pp. 191–197, Zurich, 2013.

[31] G. Tao. *Adaptive Control Design and Analysis*. John Wiley & Sons, 2003.

[32] A. J. van der Schaft. L_2-*Gain and Passivity in Nonlinear Control*. Springer, Berlin, 1999.

[33] D. Vrabie and F. L. Lewis. Neural network approach to continuous-time direct adaptive optimal control for partially unknown nonlinear systems. *Neural Networks*, 22(3):237–246, 2009.

[34] Y. Wang, B. O'Donoghue, and S. Boyd. Approximate dynamic programming via iterated Bellman inequalities. *International Journal of Robust and Nonlinear Control*, 25(10):1472–1496, July 2015.

[35] Y. Wang and S. Boyd. Performance bounds and suboptimal policies for linear stochastic control via LMIs. *International Journal of Robust and Nonlinear Control*, 21(14):1710–1728, 2011.

[36] J. Xu, L. Xie, and Y. Wang. Simultaneous stabilization and robust control of polynomial nonlinear systems using SOS techniques. *IEEE Transactions on Automatic Control*, 54(8):1892–1897, 2009.

CHAPTER 5

ROBUST ADAPTIVE DYNAMIC PROGRAMMING

In Chapters 2–4, ADP-based methodologies for continuous-time systems have been developed, under the assumption that the system order is known and the state variables are fully available for both learning and control purposes. However, this assumption may not be satisfied due to the presence of dynamic uncertainties (or unmodeled dynamics) [10], which are motivated by engineering applications in situations where the exact mathematical model of a physical system is not easy to be obtained. Of course, dynamic uncertainties also make sense for the mathematical modeling in other branches of science such as biology and economics. This problem, often formulated in the context of robust control theory, cannot be viewed as a special case of output feedback control. In addition, the ADP methods developed in the past literature may fail to guarantee not only optimality, but also the stability of the closed-loop system when dynamic uncertainty occurs. To overcome these difficulties, this chapter introduces a new concept of robust adaptive dynamic programming (RADP), a natural extension of ADP to uncertain dynamic systems (see Figure 5.1). It is worth noting that we focus on the presence of dynamic uncertainties, of which the state variables and the system order are not precisely known.

Robust Adaptive Dynamic Programming, First Edition. Yu Jiang and Zhong-Ping Jiang.
© 2017 by The Institute of Electrical and Electronics Engineers, Inc. Published 2017 by John Wiley & Sons, Inc.

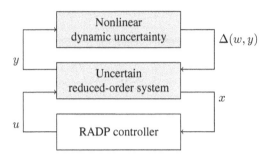

FIGURE 5.1   RADP-based online learning control for uncertain nonlinear systems.

5.1 RADP FOR PARTIALLY LINEAR COMPOSITE SYSTEMS

5.1.1 Optimality and Robustness

5.1.1.1 Systems with Matched Disturbance Input   Consider the system

$$\dot{x} = Ax + B(u + D\Delta),$$
$$y = Cx \tag{5.1}$$

where $x \in \mathbb{R}^n$ is the system state; $u \in \mathbb{R}^m$ is the control input; $A \in \mathbb{R}^{n\times n}$, $B \in \mathbb{R}^{n\times m}$, $C \in \mathbb{R}^{q\times n}$, and $D \in \mathbb{R}^{m\times p}$ are uncertain system matrices with (A, B) stabilizable and (A, C) observable; $\Delta \in \mathbb{R}^p$ is the disturbance signal. The system is said to have *matched* disturbance input, in the sense that the disturbance input $D\Delta$ is in the range of the control input.

For the above-mentioned system, the relationship between optimality and robustness for a given controller can be stated in the following lemma.

Lemma 5.1.1   *Consider system (5.1). Assume $u = -K^*x$ is the optimal control law for the disturbance-free system (5.1), that is, when $\Delta = 0$, it minimizes the following cost*

$$J_0(x_0; u) = \int_0^\infty (x^T Q x + u^T R u) d\tau \tag{5.2}$$

where Q and R are two real symmetric matrices satisfying

$$Q \geq \gamma^{-1} C^T C \text{ and } R^{-1} \geq DD^T$$

for any given constant $\gamma > 0$. Then, there exists a positive definite and radially unbounded function $V_0(x)$ such that, along the solutions of system (5.1), we have

$$\dot{V}_0 \leq -\gamma |y|^2 + |\Delta|^2 \tag{5.3}$$

Proof: By optimal control theory, we have

$$u = -K^*x = -R^{-1}B^T P^* x \tag{5.4}$$

where $P^* = P^{*T} > 0$ is the solution to the following algebraic Riccati equation (ARE)

$$PA + A^T P + Q - PBR^{-1}B^T P = 0 \tag{5.5}$$

Define $V_0 = x^T P^* x$. Then, by completing the squares, along the solutions of the closed-loop system composed of (5.1) and (5.4), we have

$$
\begin{aligned}
\dot{V}_0 &= x^T P^* (A - BK^*) x + x^T (A - BK^*)^T P^* x + 2x^T P^* BD\Delta \\
&\le -x^T Q x - x^T P^* B D D^T B^T P^* x + 2x^T P^* BD\Delta - |\Delta|^2 + |\Delta|^2 \\
&\le -x^T Q x - |D^T B^T P^* x - \Delta|^2 + |\Delta|^2 \\
&\le -\gamma^{-1} |y|^2 + |\Delta|^2
\end{aligned}
$$

The proof is complete.   ■

Clearly, the selection of the weighting matrices Q and R determines the input-to-output relationship of system (5.1). Next, we extend the methodology to systems with unmatched disturbance input.

5.1.1.2 Systems with Unmatched Disturbance Input
Consider the following system with both matched and unmatched disturbance inputs

$$\dot{x} = Ax + B(z + D\Delta), \tag{5.6}$$
$$\dot{z} = Ex + Fz + G(u + H\Delta), \tag{5.7}$$
$$y = Cx \tag{5.8}$$

where $[x^T, z^T]^T \in \mathbb{R}^{n+m}$ is the system state vector; A, B, C, D, and Δ are defined as in (5.1); $E \in \mathbb{R}^{m \times n}$, $F \in \mathbb{R}^{m \times m}$, $G \in \mathbb{R}^{m \times m}$, and $H \in \mathbb{R}^{m \times q}$ are unknown constant matrices. It is assumed that G is nonsingular. In this case, $D\Delta$, not falling in the input space, is the unmatched disturbance, while $H\Delta$ is viewed as the matched disturbance.

For the sake of optimal controller design and robustness analysis, let us perform the following state transformation:

$$\xi = z + K^* x \tag{5.9}$$

Then, with (5.7), it follows that

$$
\begin{aligned}
\dot{\xi} &= Ex + F(\xi - K^* x) + G(u + H\Delta) \\
&\quad + K^* A_c x + K^* B\xi + K^* BD\Delta \\
&= \bar{E}x + \bar{F}\xi + G(u + \bar{D}\Delta) \tag{5.10}
\end{aligned}
$$

where

$$\bar{E} = E + K^* A_c - FK^*,$$
$$\bar{F} = F + K^* B,$$
$$\bar{D} = H + G^{-1} K^* BD,$$
$$A_c = A - BK^*. \tag{5.11}$$

Next, let $S^* = S^{*T} > 0$ be the solution to the following ARE:

$$\bar{F}^T S^* + S^* \bar{F} + W - S^* G R_1^{-1} G^T S^* = 0 \tag{5.12}$$

where $W = W^T > 0$ and $R_1^{-1} > \bar{D}\bar{D}^T$.

Now, with $\Delta \equiv 0$, the system (5.6) and (5.7) can be rewritten in a compact matrix form

$$\dot{X} = A_1 X + G_1 v \tag{5.13}$$

where

$$X = \begin{bmatrix} x^T \\ \xi^T \end{bmatrix},$$
$$v = u + G^{-1}[\bar{E} + (S^*)^{-1} B^T P^*]x,$$
$$A_1 = \begin{bmatrix} A_c & B \\ -(S^*)^{-1} B^T P^* & \bar{F} \end{bmatrix},$$
$$G_1 = \begin{bmatrix} 0 \\ G \end{bmatrix}. \tag{5.14}$$

Lemma 5.1.2 *Consider the cost*

$$J_1(X(0); u) = \int_0^\infty \left(X^T Q_1 X + v^T R_1 v \right) d\tau \tag{5.15}$$

associated with system (5.13) and let

$$Q_1 = \begin{bmatrix} Q + K^{*T} R K^* & 0 \\ 0 & W \end{bmatrix},$$
$$W > 0,$$
$$R_1^{-1} \geq \bar{D}\bar{D}^T.$$

Then, under the conditions of Lemma 5.1.1, the following control policy minimizes the cost defined in (5.15) for system (5.13)

$$v = -M^* \xi = -R_1^{-1} G^T S^* \xi \tag{5.16}$$

In addition, there is a positive definite and radially unbounded function $V_1(x, \xi)$ such that, along the solutions of (5.6)–(5.8), we have

$$\dot{V}_1 \leq -\gamma^{-1}|y|^2 + 2|\Delta|^2 \tag{5.17}$$

Proof: It is easy to check that

$$P_1^* \triangleq \begin{bmatrix} P^* & 0 \\ 0 & S^* \end{bmatrix} \tag{5.18}$$

is the solution to the following ARE

$$A_1^T P_1 + P_1 A_1 + Q_1 - P_1 G_1 R_1^{-1} G_1^T P_1 = 0 \tag{5.19}$$

Therefore, by linear optimal control theory [16], the optimal control policy is

$$v = -R_1^{-1}[0 \quad G^T]P_1^* X = -M^* \xi \tag{5.20}$$

Define $V_1(x, \xi) = X^T P_1^* X$. Then, along the solutions of (5.6) and (5.7), we have

$$
\begin{aligned}
\dot{V}_1 &= \frac{d}{dt}(x^T P^* x) + \frac{d}{dt}(\xi^T S^* \xi) \\
&\leq -\gamma^{-1}|y|^2 + |\Delta|^2 + 2x^T P^* B\xi + |\Delta|^2 - 2x^T BP^* \xi \\
&= -\gamma^{-1}|y|^2 + 2|\Delta|^2
\end{aligned} \tag{5.21}
$$

The proof is thus complete.   ■

Remark 5.1.3   *It is worth noting that a repeated application of Lemma 5.1.2 yields a robust optimal controller for higher-dimensional systems with a lower-triangular structure. In other words, the resulted feedback system is input-to-output stable from the disturbance input to the output, with an arbitrarily small L_2-gain, and, when the disturbance vanishes, is optimal with respect to some integral-quadratic cost function. Also, see [6].*

5.1.1.3 Partially Linear Composite Systems with Disturbances   Now, let
us consider the following partially linear composite system with disturbances:

$$
\begin{aligned}
\dot{w} &= f(w, y), & (5.22) \\
\dot{x} &= Ax + B[z + \Delta_1(w, y)], & (5.23) \\
\dot{z} &= Ex + Fz + G[u + \Delta_2(w, y)], & (5.24) \\
y &= Cx & (5.25)
\end{aligned}
$$

where $\Delta_1(w, y) = D\Delta(w, y)$ and $\Delta_2(w, y) = H\Delta(w, y)$ are the outputs of the dynamic uncertainty; $f : \mathbb{R}^{n_w} \times \mathbb{R}^q \to \mathbb{R}^{n_w}$ and $\Delta : \mathbb{R}^{n_w'} \times \mathbb{R}^q \to \mathbb{R}^p$ are unknown locally Lipschitz functions satisfying $f(0, 0) = 0$, $\Delta(0, 0) = 0$; other variables and all the system matrices are defined the same as in (5.6)–(5.8). Notice that the class of nonlinear uncertain systems includes the popular class of partially linear composite systems introduced by Saberi, Kokotovic, and Sussmann [19].

The primary objective of this chapter is to present an online learning strategy for the design of robust adaptive suboptimal controllers that globally asymptotically stabilize the system (5.22)–(5.25) at the origin. To begin with, let us make a few assumptions on (5.22), as often required in the literature of nonlinear control design [10, 14] and [4]. See Appendix A for more details.

Assumption 5.1.4   *The w-subsystem (5.22) has strong unboundedness observability (SUO) property with zero offset [11] and is input-to-output stable (IOS) with respect to y as the input and Δ as the output [11, 21].*

Assumption 5.1.5   *There exist a continuously differentiable, positive definite, radially unbounded function $U : \mathbb{R}^{n_w} \to \mathbb{R}_+$, and a constant $c \geq 0$ such that*

$$\dot{U} = \frac{\partial U(w)}{\partial w} f(w, y) \leq -2|\Delta|^2 + c|y|^2 \tag{5.26}$$

for all $w \in \mathbb{R}^{n_w}$ and $y \in \mathbb{R}^q$.

Assumptions 5.1.4 and 5.1.5 impose the standard yet mild requirements on the dynamic uncertainty (5.22). Indeed, instead of requiring the knowledge of the dynamic uncertainty, we only assume a bound on the input-to-output gain of this subsystem. Very often, finding such a bound is much more relaxed than those assumptions appeared in the model-based nonlinear control literature that require identifying the dynamics of the subsystem.

The following theorem gives the small-gain condition for the robust asymptotic stability of the overall system (5.22)–(5.25).

Theorem 5.1.6   *Under Assumptions 5.1.4 and 5.1.5, the control policy*

$$u^* = -[(M^{*T}R_1)^{-1}(N^* + RK^*) + M^*K^*]\, x - M^*z \tag{5.27}$$

globally asymptotically stabilizes the closed-loop system comprised of (5.22)–(5.25), with $N^* = S^*\bar{E}$, if the small-gain condition holds

$$\gamma c < 1 \tag{5.28}$$

Proof: Define

$$V(x, z, w) = x^T P^* x + \xi^T S^* \xi + U(w) \tag{5.29}$$

Then, along the solutions of the closed-loop system comprised of (5.22)–(5.25) and (5.27), by completing the squares, it follows that

$$
\begin{aligned}
\dot{V} &= \frac{d}{dt}(x^T P^* x) + \frac{d}{dt}(\xi^T S^* \xi) + \dot{U} \\
&\leq -\gamma^{-1}|y|^2 + |\Delta|^2 + 2x^T P^* B\xi + |\Delta|^2 \\
&\quad -2\xi^T B^T P^* x + (c|y|^2 - 2|\Delta|^2) \\
&\leq -\gamma^{-1}(1 - c\gamma)|y|^2
\end{aligned}
$$

Therefore, we know $\lim_{t\to\infty} y(t) = 0$. By Assumption 5.1.4, all solutions of the closed-loop system are globally bounded. Moreover, a direct application of LaSalle's Invariance Principle (Theorem A.1.2, also see [12]) yields the global asymptotic stability (GAS) property of the trivial solution of the closed-loop system.

The proof is thus complete.   ∎

Also, recall that the control policy (5.27) is suboptimal, in the sense that it is optimal with respect to (5.15) only when the dynamic uncertainty is absent.

Remark 5.1.7   *It is of interest to note that Theorem 5.1.6 can be generalized to higher-dimensional systems with a lower-triangular structure, by a repeated application of backstepping and small-gain techniques in nonlinear control.*

Remark 5.1.8   *The cost function introduced here is different from the ones used in game theory [1, 23], where the policy iterations are developed based on the game algebraic Riccati equation (GARE). The existence of a solution of the GARE cannot be guaranteed when the input-output gain is arbitrarily small. Therefore, a significant advantage of our method versus the game-theoretic approach of [1, 23] is that we are able to render the gain arbitrarily small.*

5.1.2 RADP Design

We now develop an RADP scheme to approximate the robust optimal control policy (5.27). This scheme contains two learning phases. The first learning phase consists of generating sequences of estimates of the unknown matrices K^* and P^*. Then, based on the results derived from phase-one, the second learning phase further yields sequences of estimates of the unknown matrices S^*, M^*, and N^*. It is worth noticing that the knowledge of the system matrices is not required in our learning algorithm. In addition, we will analyze the robust asymptotic stability of the overall system under the approximated control policy obtained from our algorithm.

5.1.3 Phase-One Learning

First, similar to other policy iteration-based algorithms, an initial stabilizing control policy is required in the learning phases. Here, we assume there exist an initial control policy $u_0 = -K_x x - K_z z$ and a positive definite matrix $\bar{P} = \bar{P}^T$ satisfying $\bar{P} > P_1^*$, such that along the trajectories of the closed-loop system comprised of (5.22)–(5.25) and u_0, we have

$$\frac{d}{dt}[X^T \bar{P} X + U(w)] \leq -\epsilon |X|^2 \tag{5.30}$$

where $\epsilon > 0$ is a constant. Notice that this initial stabilizing control policy u_0 can be obtained using the idea of gain assignment [11].

Next, recall that we have shown in Chapter 2 that given K_0 such that $A - BK_0$ is Hurwitz, the ARE in (5.5) can be numerically solved by iteratively finding P_k and K_k from

$$0 = (A - BK_k)^T P_k + P_k (A - BK_k) + Q + K_k^T R K_k, \tag{5.31}$$

$$K_{k+1} = R^{-1} B^T P_k. \tag{5.32}$$

Now, assume all the conditions of Theorem 5.1.6 are satisfied. Let $u = u_0 + e$ be the input signal to the system (5.22)–(5.25), with u_0 the initial globally stabilizing control policy and e an exploration noise. Then, along the trajectories of (5.24), it follows that

$$x^T(t + \delta t) P_k x(t + \delta t) - x^T(t) P_k x(t)$$
$$= 2 \int_t^{t+\delta t} (z + \Delta_1 + K_k x)^T R K_{k+1} x d\tau$$
$$- \int_t^{t+\delta t} x^T (Q + K_k^T R K_k) x d\tau \tag{5.33}$$

Using Kronecker product representation, (5.33) can be rewritten as

$$x^T \otimes x^T \Big|_t^{t+\delta t} \text{vec}(P_k)$$
$$= 2 \left[\int_t^{t+\delta t} x^T \otimes (z + \Delta_1 + K_k x)^T d\tau \right] (I_n \otimes R) \text{vec}(K_{k+1})$$
$$- \left[\int_t^{t+\delta t} x^T \otimes x^T d\tau \right] \text{vec}(Q + K_k^T R K_k) \tag{5.34}$$

For any $\phi \in \mathbb{R}^{n_\phi}$, $\varphi \in \mathbb{R}^{n_\psi}$, and sufficiently large $l > 0$, we define the operators $\delta_{\phi\psi} : \mathbb{R}^{n_\phi} \times \mathbb{R}^{n_\psi} \to \mathbb{R}^{l \times n_\phi n_\psi}$ and $I_{\phi\psi} : \mathbb{R}^{n_\phi} \times \mathbb{R}^{n_\psi} \to \mathbb{R}^{l \times n_\phi n_\psi}$ such that

$$\delta_{\phi\psi} = \left[\phi \otimes \psi |_{t_1}^{t_1 + \delta t} \quad \phi \otimes \psi |_{t_2}^{t_2 + \delta t} \quad \cdots \quad \phi \otimes \psi |_{t_l}^{t_l + \delta t} \right]^T,$$

$$I_{\phi\psi} = \left[\int_{t_1}^{t_1 + \delta t} \phi \otimes \psi d\tau \quad \int_{t_2}^{t_2 + \delta t} \phi \otimes \psi d\tau \quad \cdots \quad \int_{t_l}^{t_l + \delta t} \phi \otimes \psi d\tau \right]^T$$

where $0 \leq t_1 < t_2 < \cdots < t_l$ are arbitrary constants.

Then, (5.34) implies the following matrix form of linear equations

$$\Theta_k \begin{bmatrix} \text{vec}(P_k) \\ \text{vec}(K_{k+1}) \end{bmatrix} = \Xi_k \qquad (5.35)$$

where $\Theta_k \in \mathbb{R}^{l \times n(n+m)}$ and $\Xi_k \in \mathbb{R}^l$ are defined as

$$\Theta_k = \left[\delta_{xx} - 2I_{xx} \left(I_n \otimes K_k^T R \right) - 2(I_{xz} + I_{x\Delta_1})(I_n \otimes R) \right],$$
$$\Xi_k = -I_{xx}\text{vec}(Q + K_k^T R K_k).$$

Given K_k such that $A - BK_k$ is Hurwitz, if there is a unique pair of matrices (P_k, K_{k+1}), with $P_k = P_k^T$, satisfying (5.35), we are able to replace (5.31) and (5.32) with (5.35). In this way, the iterative process does not need the knowledge of A and B.

Next, we approximate the matrices S^*, M^*, and N^*, which also appear in (5.27).

5.1.4 Phase-Two Learning

For the matrix $K_k \in \mathbb{R}^{m \times n}$ obtained from phase-one learning, let us define

$$\hat{\xi} = z + K_k x \qquad (5.36)$$

Then,

$$\dot{\hat{\xi}} = E_k x + F_k \hat{\xi} + G(u + \Delta_2) + K_k B \Delta_1 \qquad (5.37)$$

where $E_k = E + K_k(A - BK_k) - FK_k$ and $F_k = F + K_k B$.

Similarly as in phase-one learning, we seek the online implementation of the following iterative equations:

$$0 = S_{k,j} F_{k,j} + F_{k,j}^T S_{k,j} + W + M_{k,j}^T R_1 M_{k,j} \qquad (5.38)$$
$$M_{k,j+1} = R_1^{-1} G^T S_{k,j} \qquad (5.39)$$

where $F_{k,j} = F_k - GM_{k,j}$, and we assume there exists $M_{k,0}$ such that $F_{k,0}$ is Hurwitz.

Now, along the solutions of (5.37), we have

$$\hat{\xi}^T S_{k,j} \hat{\xi} \Big|_t^{t+T} = - \int_t^{t+\delta t} \hat{\xi}^T \left(W + M_{k,j}^T R_1 M_{k,j} \right) \hat{\xi} d\tau$$
$$+ 2 \int_t^{t+\delta t} (\hat{u} + M_{k,j}\hat{\xi})^T R_1 M_{k,j+1} \hat{\xi} d\tau$$
$$+ 2 \int_t^{t+\delta t} \hat{\xi}^T N_{k,j} x d\tau + 2 \int_t^{t+\delta t} \Delta_1^T L_{k,j} \hat{\xi} d\tau$$

where $\hat{u} = u + \Delta_2$, $N_{k,j} = S_{k,j} E_k$, and $L_{k,j} = B^T K_k^T S_{k,j}$.

Then, we obtain the following linear equations that can be used to approximate the solution to the ARE (5.12)

$$\Phi_{k,j}\text{vec}([\, S_{k,j} \quad M_{k,j+1} \quad N_{k,j} \quad L_{k,j} \,]) = \Psi_{k,j} \tag{5.40}$$

where $\Phi_{k,j} \in \mathbb{R}^{l \times m(n+m)}$ and $\Psi_{k,j} \in \mathbb{R}^l$ are defined as

$$\Phi_{k,j} = \left[\delta_{\hat{\xi}\hat{\xi}}, -2I_{\hat{\xi}\hat{\xi}} \left(I_m \otimes M_{k,j}^T R_1 \right) - 2I_{\hat{\xi}\hat{u}}(I_m \otimes R_1), -2I_{x\hat{\xi}}, -2I_{\hat{\xi}\Delta_1} \right],$$
$$\Psi_{k,j} = -I_{\hat{\xi}\hat{\xi}}\text{vec}(W_k).$$

Notice that $\delta_{\hat{\xi}\hat{\xi}}, I_{\hat{\xi}\hat{u}}, I_{\hat{\xi}\hat{\xi}}, I_{\hat{\xi}\Delta_1} \in \mathbb{R}^{l \times m^2}$, $I_{x\hat{\xi}} \in \mathbb{R}^{l \times nm}$ can be obtained by

$$\delta_{\hat{\xi}\hat{\xi}} = \delta_{zz} + 2\delta_{xz}\left(K_k^T \otimes I_m \right) + \delta_{xx}\left(K_k^T \otimes K_k^T \right),$$
$$I_{\hat{\xi}\hat{u}} = I_{z\hat{u}} + I_{x\hat{u}}\left(K_k^T \otimes I_m \right),$$
$$I_{\hat{\xi}\hat{\xi}} = I_{zz} + 2I_{xz}\left(K_k^T \otimes I_m \right) + I_{xx}\left(K_k^T \otimes K_k^T \right),$$
$$I_{x\hat{\xi}} = I_{xz} + I_{xx}\left(I_n \otimes K_k^T \right),$$
$$I_{\hat{\xi}\Delta_1} = I_{x\Delta_1}\left(K_k^T \otimes I_m \right) + I_{z\Delta_1}.$$

Clearly, (5.40) does not rely on the knowledge of E, F, or G.

The RADP scheme can thus be summarized in the following algorithm.

Algorithm 5.1.9   *Off-policy RADP algorithm for nonlinear uncertain system (5.22)–(5.25)*

(1) *Initialization:*
   Find an initial stabilizing control policy $u = u_0 + e$ for the system (5.22)–(5.25). Let $k = 0$ and $j = 0$. Set Q, R, W, and R_1, according to Lemma 5.1.1, Lemma 5.1.2, and Theorem 5.1.6. Select a sufficiently small constant $\epsilon > 0$.

(2) *Online data collection:*
   Apply the initial control policy $u = u_0$ to the system (5.22)–(5.25), until the rank condition in Lemma 5.1.10 is satisfied.

(3) *Phase-one learning:*
   Solve P_k, K_{k+1} from (5.35). Repeat this step with $k \leftarrow k+1$, until $|P_k - P_{k-1}| < \epsilon$.

(4) *Phase-two learning:*
   Solve $S_{k,j}$, $M_{k,j+1}$, $N_{k,j}$, and $L_{k,j}$ from (5.40). Let $j \leftarrow j+1$. Repeat this step with $j \leftarrow j+1$, until $|S_{k,j} - S_{k,j-1}| < \epsilon$.

(5) *Exploitation:*
   Use

$$\tilde{u} = -\left[\left(M_{k,j}^T R_1 \right)^{-1} (N_{k,j} + RK_k) + M_{k,j}K_k \right] x - M_{k,j}z \tag{5.41}$$

   as the approximate control input.

By continuity, the control policy (5.41) stabilizes (5.22)–(5.25), if it converges to (5.27) and $\epsilon > 0$ is sufficiently small. The convergence property is analyzed in Section 5.1.5.

5.1.5 Convergence Analysis

Lemma 5.1.10   *Suppose A_k and F_{kj} are Hurwitz and there exists an integer $l_0 > 0$, such that the following holds for all $l \geq l_0$:*

$$\text{rank}([\, I_{xx} \quad I_{xz} \quad I_{zz} \quad I_{x\hat{u}} \quad I_{z\hat{u}} \quad I_{x\Delta_1} \quad I_{z\Delta_1} \,])$$
$$= \frac{n(n+1)}{2} + \frac{m(m+1)}{2} + 3mn + 2m^2 \tag{5.42}$$

Then,

(1) *there exist unique $P_k = P_k^T$ and K_{k+1} satisfying (5.35), and*
(2) *there exist unique $S_{kj} = S_{kj}^T$, $M_{k,j+1}$, N_{kj}, L_{kj} satisfying (5.40).*

Proof: The proof of (1) has been given in the previous section, and is restated here for the reader's convenience. Actually, we only need to show that, given any constant matrices $P = P^T \in \mathbb{R}^{n \times n}$ and $K \in \mathbb{R}^{m \times n}$, if

$$\Theta_k \begin{bmatrix} \text{vec}(P) \\ \text{vec}(K) \end{bmatrix} = 0, \tag{5.43}$$

we will have $P = 0$ and $K = 0$.
By definition, we have

$$\Theta_k \begin{bmatrix} \text{vec}(P) \\ \text{vec}(K) \end{bmatrix} = I_{xx}\text{vec}(Y) + 2(I_{xz} + I_{x\Delta_1})\text{vec}(Z) \tag{5.44}$$

where

$$Y = A_k^T P + P A_k + K_k^T (B^T P - RK) + (PB - K^T R)K_k, \tag{5.45}$$
$$Z = B^T P - RK. \tag{5.46}$$

Since $Y = Y^T$, there are at most $\frac{1}{2}n(n+1) + mn$ distinct entries in both Y and Z. On the other hand, under the rank condition in Lemma 5.1.10, we have

$$\text{rank}([\, I_{xx} \quad I_{xz} + I_{x\Delta_1} \,]) \geq \text{rank}([I_{xx}, I_{xz}, I_{zz}, I_{x\hat{u}}, I_{z\hat{u}}, I_{x\Delta_1}, I_{z\Delta_1}])$$
$$- 2mn - \frac{1}{2}m(m+1) - 2m^2$$
$$= \frac{1}{2}n(n+1) + mn$$

implying that the $\frac{1}{2}n(n+1) + mn$ distinct columns in the matrix $[I_{xx} \quad I_{xz} + I_{x\Delta_1}]$ are linearly independent. Hence, $Y = Y^T = 0$ and $Z = 0$.

Finally, since A_k is Hurwitz for each $k \in \mathbb{Z}_+$, the only matrices $P = P^T$ and K simultaneously satisfying (5.45) and (5.46) are $P = 0$ and $K = 0$.

Now we prove (2). Similarly, suppose there exist some constant matrices $S, M, L \in \mathbb{R}^{m \times m}$ with $S = S^T$, and $N \in \mathbb{R}^{m \times n}$ satisfying

$$\Phi_{k,j} \text{vec}([S \quad M \quad N \quad L]) = 0$$

Then, we have

$$
\begin{aligned}
0 = {} & I_{\hat{\xi}\hat{\xi}}\text{vec}\left[SF_{k,j} + F_{k,j}^T S + M_{k,j}^T(G^T S - R_1 M)(SG - M^T R_1)M_{k,j}\right] \\
& + I_{\hat{\xi}\hat{u}}2\text{vec}(G^T S - R_1 M) + I_{x\hat{\xi}}2\text{vec}(SE_k - N) \\
& + I_{\hat{\xi}\Delta_1}2\text{vec}(B^T K_k^T S - L)
\end{aligned}
$$

By definition, it holds

$$[I_{xx}, I_{\hat{\xi}\hat{\xi}}, I_{x\hat{\xi}}, I_{\hat{\xi}\hat{u}}, I_{x\hat{u}}, I_{x\Delta_1}, I_{\hat{\xi}\Delta_1}] = [I_{xx}, I_{xz}, I_{zz}, I_{x\hat{u}}, I_{z\hat{u}}, I_{x\Delta_1}, I_{z\Delta_1}] T_n$$

where T_n is a nonsingular matrix. Therefore,

$$
\begin{aligned}
& \frac{1}{2}m(m+1) + 2m^2 + mn \\
& \geq \text{rank}([I_{\hat{\xi}\hat{\xi}} \quad I_{\hat{\xi}\hat{u}} \quad I_{x\hat{\xi}} \quad I_{\hat{\xi}\Delta_1}]) \\
& \geq \text{rank}([I_{xx}, I_{\hat{\xi}\hat{\xi}}, I_{x\hat{\xi}}, I_{\hat{\xi}\hat{u}}, I_{x\hat{u}}, I_{x\Delta_1}, I_{\hat{\xi}\Delta_1}]) - \frac{1}{2}n(n+1) - 2mn \\
& = \text{rank}([I_{xx}, I_{xz}, I_{zz}, I_{x\hat{u}}, I_{z\hat{u}}, I_{x\Delta_1}, I_{z\Delta_1}]) - \frac{1}{2}n(n+1) - 2mn \\
& = \frac{1}{2}m(m+1) + 2m^2 + mn
\end{aligned}
$$

Following the same reasoning from (5.44) to (5.46), we obtain

$$0 = SF_{k,j} + F_{k,j}^T S + M_{k,j}^T(G^T S - R_1 M) + (SG - M^T R_1)M_{k,j}, \qquad (5.47)$$

$$0 = G^T S - R_1 M, \qquad (5.48)$$

$$0 = SE - N, \qquad (5.49)$$

$$0 = BK_k S - L \qquad (5.50)$$

where $[S, E, M, L] = 0$ is the only possible solution. ∎

5.2 RADP FOR NONLINEAR SYSTEMS

In this section, we introduce the robust redesign technique to achieve RADP for nonlinear systems. To begin with, let us consider the nonlinear system with dynamic uncertainties as follows:

$$\dot{w} = q(w, x) \tag{5.51}$$

$$\dot{x} = f(x) + g(x)[u + \Delta(w, x)] \tag{5.52}$$

where $x \in \mathbb{R}^n$ is the system state, $w \in \mathbb{R}^{n_w}$ is the state of the dynamic uncertainty, $u \in \mathbb{R}^m$ is the control input, $q : \mathbb{R}^{n_w} \times \mathbb{R}^n \to \mathbb{R}^{n_w}$ and $\Delta : \mathbb{R}^{n_w} \times \mathbb{R}^n \to \mathbb{R}^m$, are unknown locally Lipschitz functions, $f : \mathbb{R}^n \to \mathbb{R}^n$ and $g : \mathbb{R}^n \to \mathbb{R}^{n \times m}$ are uncertain polynomials with $f(0) = 0$.

We call the following system the *reduced-order* system, in the sense that (5.51)–(5.52) would reduce to it in the absence of the dynamic uncertainty.

$$\dot{x} = f(x) + g(x)u \tag{5.53}$$

In the previous chapters, we have developed ADP methods to optimize the reduced-order system by minimizing the following cost.

$$J(x_0; u) = \int_0^\infty [q(x) + u^T R u] dt \tag{5.54}$$

Apparently, in the presence of the dynamic uncertainty, the approximate optimal control policies developed for the reduced-order system may fail to stabilize the overall system (5.51) and (5.52). To overcome this difficulty, we redesign the approximate optimal control policy proposed in previous chapters, such that they can achieve global stabilization for the overall system (5.51) and (5.52).

5.2.1 Robust Redesign

To begin with, let us assume in (5.54), the weighting function $q(x)$ is selected such that $q(x) = q_0(x) + \epsilon|x|^2$, with $q_0(x)$ a positive definite function and $\epsilon > 0$ a constant, and R is selected as a real symmetric and positive definite matrix. Next, we introduce two Assumptions on the system (5.51) and (5.52).

Assumption 5.2.1 *Consider the system comprised of (5.51) and (5.52). There exist functions $\underline{\lambda}, \bar{\lambda} \in \mathcal{K}_\infty$, $\kappa_1, \kappa_2, \kappa_3 \in \mathcal{K}$, and positive definite functions W and κ_4, such that for all $w \in \mathbb{R}^p$ and $x \in \mathbb{R}^n$, we have*

$$\underline{\lambda}(|w|) \le W(w) \le \bar{\lambda}(|w|), \tag{5.55}$$

$$|\Delta(w, x)| \le \kappa_1(|w|) + \kappa_2(|x|), \tag{5.56}$$

together with the following implication:

$$W(w) \geq \kappa_3(|x|) \Rightarrow \nabla W(w)^T q(w, x) \leq -\kappa_4(w). \tag{5.57}$$

Notice that Assumption 5.2.1 implies that the w-subsystem (5.51) is input-to-state stable (ISS) [20, 22] with respect to x as the input. See Appendix A for a detailed review of the ISS property.

Assumption 5.2.2   *There exist a value function $V_i \in \mathcal{P}$ and a control policy u_i satisfying*

$$\nabla V_i^T(x) \left(f(x) + g(x) u_i \right) + q(x) + u_i^T R u_i \leq 0, \forall x_0 \in \mathbb{R}^n. \tag{5.58}$$

In addition, there exist $\underline{\alpha}, \bar{\alpha} \in \mathcal{K}_\infty$, such that the following inequalities hold:

$$\underline{\alpha}(|x|) \leq V_i(x) \leq \bar{\alpha}(|x|), \forall x_0 \in \mathbb{R}^n \tag{5.59}$$

Under Assumptions 5.2.1 and 5.2.2, the control policy can be redesigned as

$$u_{r,i} = \rho^2(|x|^2) u_i \tag{5.60}$$

where $\rho(\cdot)$ is a smooth and nondecreasing function with $\rho(s) \geq 1, \forall s > 0$.

Theorem 5.2.3   *Under Assumptions 5.2.1 and 5.2.2, the closed-loop system comprised of (5.51), (5.52), and*

$$u = u_{r,i} + e \tag{5.61}$$

is ISS with respect to e as the input, if the following gain condition holds:

$$\gamma > \kappa_1 \circ \underline{\lambda}^{-1} \circ \kappa_3 \circ \underline{\alpha}^{-1} \circ \bar{\alpha} + \kappa_2, \tag{5.62}$$

where $\gamma \in \mathcal{K}_\infty$ is defined by

$$\gamma(s) = \epsilon s \sqrt{\frac{\frac{1}{4} + \frac{1}{2} \rho^2(s^2)}{\lambda_{\min}(R)}}. \tag{5.63}$$

Proof: Let $\chi_1 = \kappa_3 \circ \underline{\alpha}^{-1}$. Then, under Assumption 5.2.1, we immediately have the following implications

$$W(w) \geq \chi_1(V_i(x))$$
$$\Rightarrow W(w) \geq \kappa_3(\underline{\alpha}^{-1}(V_i(x))) \geq \kappa_3(|x|)$$
$$\Rightarrow \nabla W^T(w) q(w, x) \leq -\kappa_4(w) \tag{5.64}$$

Define $\tilde{\rho}(x) = \sqrt{\dfrac{\frac{1}{4} + \frac{1}{2}\rho^2(|x|^2)}{\lambda_{\min}(R)}}$. Then, along the solutions of system (5.52), it

follows that

$$
\begin{aligned}
&\nabla V_i^T[f + g(u_{r,i} + e + \Delta)] \\
&\leq -q(x) - u_i^T R u_i + \nabla V_i^T g[(\rho^2(|x|^2) - 1)u_i + \Delta + e] \\
&\leq -q(x) - \tilde{\rho}^2 |g^T \nabla V_i|^2 + \nabla V_i^T g(\Delta + e) \\
&\leq -q(x) - \left| \tilde{\rho} g^T \nabla V_i - \frac{1}{2}\tilde{\rho}^{-1}\Delta \right|^2 + \frac{1}{4}\tilde{\rho}^{-2}|\Delta + e|^2 \\
&\leq -q_0(x) - e^2|x|^2 + \tilde{\rho}^{-2}\max\{|\Delta|^2, |e|^2\} \\
&\leq -q_0(x) - \tilde{\rho}^{-2}(\gamma^2 - \max\{|\Delta|^2, |e|^2\})
\end{aligned}
$$

Hence, by defining $\chi_2 = \bar{\alpha} \circ (\gamma - \kappa_2)^{-1} \circ \kappa_1 \circ \underline{\lambda}^{-1}$, it follows that

$$
\begin{aligned}
&V_i(x) \geq \max\{\chi_2(W(w)), \bar{\alpha} \circ (\gamma - \kappa_2)^{-1}(|e|)\} \\
&\Leftrightarrow V_i(x) \geq \bar{\alpha} \circ (\gamma - \kappa_2)^{-1} \circ \max\{\kappa_1 \circ \underline{\lambda}^{-1}(W(w)), |e|\} \\
&\Rightarrow (\gamma - \kappa_2) \circ \bar{\alpha}^{-1}(V_i(x)) \geq \max\{\kappa_1 \circ \underline{\lambda}^{-1}(W(w)), |e|\} \\
&\Rightarrow \gamma(|x|) - \kappa_2(|x|) \geq \max\{\kappa_1 \circ \underline{\lambda}^{-1}(W(w)), |e|\} \\
&\Rightarrow \gamma(|x|) - \kappa_2(|x|) \geq \max\{\kappa_1(|w|), |e|\} \\
&\Rightarrow \gamma(|x|) \geq \max\{|\Delta(w, x)|, |e|\} \\
&\Rightarrow \nabla V_i^T[f + g(u_{r,i+1} + e + \Delta)] \leq -q_0(x) \quad\quad (5.65)
\end{aligned}
$$

Finally, by the gain condition, we have

$$
\begin{aligned}
&\gamma > \kappa_1 \circ \underline{\lambda}^{-1} \circ \kappa_3 \circ \underline{\alpha}^{-1} \circ \bar{\alpha} + \kappa_2 \\
&\Rightarrow Id > (\gamma - \kappa_2)^{-1} \circ \kappa_1 \circ \underline{\lambda}^{-1} \circ \kappa_3 \circ \underline{\alpha}^{-1} \circ \bar{\alpha} \\
&\Rightarrow Id > \bar{\alpha} \circ (\gamma - \kappa_2)^{-1} \circ \kappa_1 \circ \underline{\lambda}^{-1} \circ \kappa_3 \circ \underline{\alpha}^{-1} \\
&\Rightarrow Id > \chi_2 \circ \chi_1 \quad\quad (5.66)
\end{aligned}
$$

The proof is thus completed by the small-gain theorem (Theorem A.2.4, also see [9]). ∎

5.2.2 Robust Policy Iteration and Online Learning via RADP

The robust redesign can be easily incorporated into the conventional policy iteration (see Algorithm 3.1.3). For the reader's convenience, we rewrite the two main assumptions introduced in Chapter 3 in the following.

Assumption 5.2.4   *There exists $V^* \in P$, such that the HJB equation holds*

$$0 = \nabla V^{*T}(x)f(x) + q(x) - \frac{1}{4}\nabla V^{*T}(x)g(x)R^{-1}(x)g^T(x)\nabla V^*(x) \qquad (5.67)$$

Assumption 5.2.5   *There exists a feedback control policy $u_0 : \mathbb{R}^n \rightarrow \mathbb{R}^m$ that globally asymptotically stabilizes the system (5.53) at the origin under a finite cost (5.54).*

Algorithm 5.2.6   *Robust policy iteration algorithm*

(1) *Initialization:*
   Find a control policy $u = u_0(x)$ that satisfies Assumption 5.2.5.
(2) *Robust redesign:*
   Robustify the control policy to

$$u_{r,i} = \rho^2(|x|^2)u_i \qquad (5.68)$$

(3) *Policy evaluation:*
   For $i = 0, 1, ...$, solve for the cost function $V_i(x) \in C^1$, with $V_i(0) = 0$, from the following partial differential equation:

$$\nabla V_i^T(x)[f(x) + g(x)u_i] + r(x, u_i) = 0 \qquad (5.69)$$

(4) *Policy improvement:*
   Update the control policy by

$$u_{i+1}(x) = -\frac{1}{2}R^{-1}(x)g^T(x)\nabla V_i(x) \qquad (5.70)$$

Properties of Algorithm 5.2.6 are summarized in the following Corollary, which results directly from Theorem 3.1.4 and Theorem 5.2.3.

Corollary 5.2.7   *Suppose Assumptions 5.2.1, 5.2.4, and 5.2.5 hold, and the solution $V_i(x) \in C^1$ satisfying (5.69) exists, for $i = 1, 2, ...$ Let $V_i(x)$ and $u_{i+1}(x)$ be the functions obtained from (5.69) and (5.70). Then, the following properties hold, $\forall i = 0, 1, ...$*

(1) $V^*(x) \leq V_{i+1}(x) \leq V_i(x)$, $\forall x \in \mathbb{R}^n$;
(2) $u_{r,i}$ *is globally asymptotically stabilizing for the system (5.51)–(5.52);*
(3) *if $\lim_{i \to \infty} V_i(x) \in C^1$, then*

$$\lim_{i \to \infty} V_i(x) = V^*(x), \qquad (5.71)$$

$$\lim_{i \to \infty} u_i(x) = u^*(x), \qquad (5.72)$$

$$\lim_{i \to \infty} u_{r,i}(x) = u_r^*(x), \qquad (5.73)$$

for any $x \in \mathbb{R}^n$.

The control policy u_r^* defined in (5.73) is denoted as the robust optimal control policy for system (5.51) and (5.52).

Algorithm 5.2.6 can be implemented online by following the same procedure as described in Chapter 3. Indeed, along the solutions of the system (5.52) and (5.61), it follows that

$$
\begin{aligned}
\dot{V}_i &= \nabla V_i^T(x)[f(x) + g(x)u_i + g(x)\tilde{e}_i] \\
&= -q(x) - u_i^T R u_i + \nabla V_i^T(x)g(x)\tilde{e}_i \\
&= -q(x) - u_i^T R u_i - 2u_{i+1}^T R \tilde{e}_i \qquad (5.74)
\end{aligned}
$$

where $\tilde{e} = (\rho^2(|x|^2) - 1)u_i + \Delta + e$.

Integrating both sides of (5.74) on any time interval $[t, t + \delta t]$, it follows that

$$
V_i(x(t + \delta t)) - V_i(x(t)) = -\int_t^{t+\delta t} \left[q(x) + u_i^T R u_i + 2u_{i+1}^T R \tilde{e}_i\right] d\tau \qquad (5.75)
$$

Thus, given u_0, the methodology developed in Chapter 3 can be performed on (5.75) to achieve online approximation of the value function V_i and the control policy u_{i+1}.

5.2.3 SOS-Based Robust Policy Iteration and Global RADP

When the reduced-order system is polynomial in x, that is, Assumption 5.2.8 is satisfied, we can also incorporate the robust redesign into the SOS-based policy iteration as introduced in Algorithm 4.3.2.

Assumption 5.2.8   *There exist smooth mappings $V_0 : \mathbb{R}^n \to \mathbb{R}$ and $u_1 : \mathbb{R}^n \to \mathbb{R}^m$, such that $V_0 \in \mathbb{R}[x]_{2,2r} \cap P$ and $\mathcal{L}(V_0, u_1)$ is SOS.*

Algorithm 5.2.9   *SOS-based robust policy iteration*

(1) *Initialization:*
   Find V_0 and u_1 such that Assumption 5.2.8 is satisfied.
(2) *Robust redesign:*
   Robustify the control policy to

$$
u_{r,i} = \rho^2(|x|^2)u_i \qquad (5.76)
$$

(3) *Policy evaluation:*
   For $i = 1, 2, \ldots$, solve for an optimal solution p_i to the following optimization problem:

$$
\min_p \quad \int_\Omega V(x)dx \qquad (5.77)
$$

$$
\text{s.t.} \quad \mathcal{L}(V, u_i) \quad \text{is} \quad \text{SOS} \qquad (5.78)
$$

$$
V_{i-1} - V \quad \text{is} \quad \text{SOS} \qquad (5.79)
$$

where $V = p^T[x]_{2,2r}$. Then, denote $V_i = p_i^T[x]_{2,2r}$.

(4) *Policy improvement:*
   Update the control policy by

$$u_{i+1} = -\frac{1}{2}R^{-1}g^T\nabla V_i. \tag{5.80}$$

Then, go to Step (2) with i replaced by i + 1.

Corollary 5.2.10   *Under Assumptions 5.2.1 and 5.2.8. The following properties hold for Algorithm 5.2.9.*

(1) *The optimization problem (5.77)–(5.79) has a nonempty feasible set.*
(2) $V_i \in P$ *and* $V_i(x) \le V_{i-1}$, *for* $i = 1, 2, \ldots$
(3) *The closed-loop system comprised of (5.51), (5.52), and (5.82) is globally asymptotically stable at the origin, if the gain condition (5.62) holds.*

Algorithm 5.2.9 can be implemented in an identical fashion as described in Chapter 4. Indeed, along the solutions of (5.51)–(5.52), we have

$$
\begin{aligned}
\dot{V} &= \nabla V^T(f + gu_{r,i}) \\
&= \nabla V^T(f + gu_i) + \nabla V^T g\tilde{e} \\
&= -r(x, u_i) - \mathcal{L}(V, u_i) + \nabla V^T g\tilde{e} \\
&= -r(x, u_i) - \mathcal{L}(V, u_i) + 2\left(\frac{1}{2}R^{-1}g^T\nabla V\right)^T R\tilde{e} \\
&= -r(x, u_i) - \iota(p, K_i)^T[x]_{2,2d} - 2[x]_{1,d}^T\kappa(p)^T R\tilde{e}
\end{aligned}
\tag{5.81}
$$

where $\tilde{e} = (\rho^2(|x|^2) - 1)u_i + \Delta + e$.

Therefore, we can redefine the data matrices as follows:

$$\bar{\sigma}_e = -[\bar{\sigma}^T \quad 2\sigma^T \otimes \tilde{e}^T R]^T \in \mathbb{R}^{l_1+ml},$$

$$\bar{\Phi}_i = \left[\int_{t_{0,i}}^{t_{1,i}} \bar{\sigma}_e dt \quad \int_{t_{1,i}}^{t_{2,i}} \bar{\sigma}_e dt \quad \cdots \quad \int_{t_{q_i-1,i}}^{t_{q_i,i}} \bar{\sigma}_e dt\right]^T \in \mathbb{R}^{q_i \times (l_1+ml)},$$

$$\bar{\Xi}_i = \left[\int_{t_{0,i}}^{t_{1,i}} r(x, u_i)dt \quad \int_{t_{1,i}}^{t_{2,i}} r(x, u_i)dt \quad \cdots \quad \int_{t_{q_i-1,i}}^{t_{q_i,i}} r(x, u_i)dt\right]^T \in \mathbb{R}^{q_i},$$

$$\bar{\Theta}_i = \left[\bar{\phi}(x)|_{t_{0,i}}^{t_{1,i}} \quad \bar{\phi}(x)|_{t_{1,i}}^{t_{2,i}} \quad \cdots \quad \bar{\phi}(x)|_{t_{q_i-1,i}}^{t_{q_i,i}}\right]^T \in \mathbb{R}^{q_i \times N_1}.$$

The online global robust adaptive dynamic programming algorithm is thus given below.

Algorithm 5.2.11   *Online global robust adaptive dynamic programming algorithm*

(1) *Initialization:*
   Find V_0 and u_1 such that Assumption 5.2.8 is satisfied.

(2) *Robust redesign:*
   Robustify the control policy to

$$u_{r,i} = \rho^2(|x|^2)u_i \tag{5.82}$$

(3) *Collect online data:*
   Apply $u = u_{r,i} + e$ to the system and compute the data matrices $\tilde{\Phi}_i$, $\tilde{\Xi}_i$, and $\tilde{\Theta}_i$, until $\tilde{\Phi}_i$ is of full-column rank.
(4) *Policy evaluation and improvement:*
   Find an optimal solution (p_i, K_{i+1}) to the following SOS program

$$\min_{p,K_p} c^T p \tag{5.83}$$

$$\text{s.t. } \tilde{\Phi}_i \begin{bmatrix} l_p \\ \text{vec}(K_p) \end{bmatrix} = \tilde{\Xi}_i + \tilde{\Theta}_i p \tag{5.84}$$

$$l_p^T[x]_{2,2d} \text{ is SOS} \tag{5.85}$$

$$(p_{i-1} - p)^T[x]_{2,2r} \text{ is SOS} \tag{5.86}$$

where $c = \int_\Omega [x]_{2,2r} dx$, with $\Omega \subset \mathbb{R}^n$ a compact set. Then, denote $V_i = p_i^T[x]_{2,2r}$, $u_{i+1} = K_{i+1}[x]_{1,d}$, and go to Step (2) with $i \leftarrow i + 1$.

5.3 APPLICATIONS

5.3.1 Application to Synchronous Generators

The power system considered in this chapter is an interconnection of two synchronous generators described by [15] (see Figure 5.2):

$$\Delta \dot{\delta}_i = \Delta \omega_i, \tag{5.87}$$

$$\Delta \dot{\omega}_i = -\frac{D}{2H_i}\Delta \omega + \frac{\omega_0}{2H_i}(\Delta P_{mi} + \Delta P_{ei}), \tag{5.88}$$

$$\Delta \dot{P}_{mi} = \frac{1}{T_i}(-\Delta P_{mi} - k_i \Delta \omega_i + u_i), i = 1, 2 \tag{5.89}$$

where, for the *i*th generator, $\Delta \delta_i$, $\Delta \omega_i$, ΔP_{mi}, and ΔP_{ei} are the deviations of rotor angle, relative rotor speed, mechanical input power, and active power, respectively. The control signal u_i represents deviation of the valve opening. H_i, D_i, ω_0, k_i, and T_i are constant system parameters.

The active power ΔP_{ei} is defined as

$$\Delta P_{e1} = -\frac{E_1 E_2}{X}[\sin(\delta_1 - \delta_2) - \sin(\delta_{10} - \delta_{20})] \tag{5.90}$$

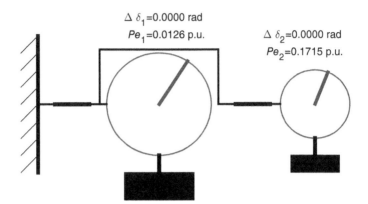

FIGURE 5.2   Illustration of a two-machine power system at steady state.

and $\Delta P_{e2} = -\Delta P_{e1}$, where δ_{10} and δ_{20} are the steady-state angles of the first and second generators. The second synchronous generator is treated as the dynamic uncertainty, and it has a fixed controller $u_2 = -a_1 \Delta \delta_2 - a_2 \Delta \omega_2 - a_3 \Delta P_{m2}$, with a_1, a_2, and a_3 its feedback gains.

Our goal is to design a robust optimal control policy u_1 for the interconnected power system. For simulation purpose, the parameters are specified as follows: $D_1 = 1$, $H_1 = 3$, $\omega_0 = 314.159$ rad/s, $T_1 = 5s$, $\delta_{10} = 2$ rad, $D_2 = 1$, $T_2 = 5$, $X = 15$, $k_2 = 0$, $H_2 = 3$, $a_1 = 0.2236$, $a_2 = -0.2487$, $a_3 = -7.8992$. Weighting matrices are $Q = \text{diag}(5, 0.0001)$, $R = 1$, $W = 0.01$, and $R_1 = 100$. The exploration noise we employed for this simulation is the sum of sinusoidal functions with different frequencies.

In the simulation, two generators were operated on their steady states from $t = 0s$ to $t = 1s$. An impulse disturbance on the load was simulated at $t = 1s$, and the overall system started to oscillate. The RADP algorithm was applied to the first generator from $t = 2s$ to $t = 3s$. Convergence is attained after 6 iterations of phase-one learning followed by 10 iterations of phase-two learning, when the stopping criteria $|P_k - P_{k-1}| \leq 10^{-6}$ and $|S_{k,j} - S_{k,j-1}| \leq 10^{-6}$ are both satisfied. The linear control policy formulated after the RADP algorithm is as follows:

$$\tilde{u}_1 = -256.9324 \Delta \delta_1 - 44.4652 \Delta \omega_1 - 153.1976 \Delta P_{m1}$$

The ideal robust optimal control policy is given for comparison as follows:

$$u_1^* = -259.9324 \Delta \delta_1 - 44.1761 \Delta \omega_1 - 153.1983 \Delta P_{m1}$$

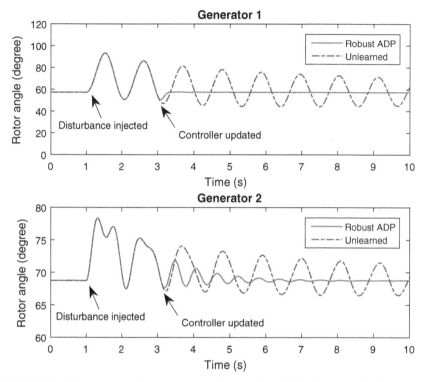

FIGURE 5.3 Trajectories of the rotor angles. *Source*: Jiang, 2013. Reproduced with permission of IEEE.

The new control policy for Generator 1 is applied from $t = 3s$ to the end of the simulation. In Figures 5.3 and 5.4, it can be seen that oscillation has been significantly reduced after RADP-based online learning.

5.3.2 Jet Engine Surge and Stall Dynamics

Consider the following system, which is inspired by the jet engine surge and stall dynamics in [13, 18]

$$\dot{r} = -\sigma r^2 - \sigma r(2\phi + \phi^2) \qquad (5.91)$$
$$\dot{\phi} = -a\phi^2 - b\phi^3 - (u + 3r\phi + 3r) \qquad (5.92)$$

where $r > 0$ is the normalized rotating stall amplitude, ϕ is the deviation of the scaled annulus-averaged flow, u is the deviation of the plenum pressure rise and is treated as the control input, $\sigma \in [0.2, 0.5]$, $a \in [1.2, 1.6]$, and $b \in [0.3, 0.7]$ are uncertain constants.

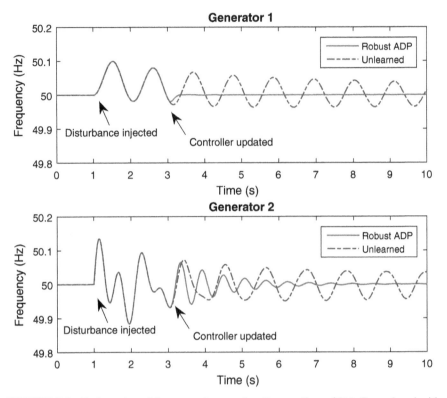

FIGURE 5.4   Trajectories of the power frequencies. *Source*: Jiang, 2013. Reproduced with permission of IEEE.

In this example, we assume the variable r is not available for real-time feedback control due to a 0.2s time delay in measuring it. Hence, the objective is to find a control policy that only relies on ϕ.

The cost function we used here is

$$J = \int_0^\infty (5\phi^2 + u^2)dt \tag{5.93}$$

and an initial control policy is chosen as

$$u_{r,1} = -\frac{1}{2}\rho^2(\phi^2)(2x - 1.4x^2 - 0.45x^3) \tag{5.94}$$

with $\rho(s) = \sqrt{2}$.

Only for the purpose of simulation, we set $\sigma = 0.3$, $a = 1.5$, and $b = 0.5$. The control policy is updated every 0.25s until the convergence criterion, $|p_i - p_{i-1}| < 0.1$ is satisfied. The simulation results are provided in Figures 5.5–5.8. It can be seen that the system performance has been improved via online learning.

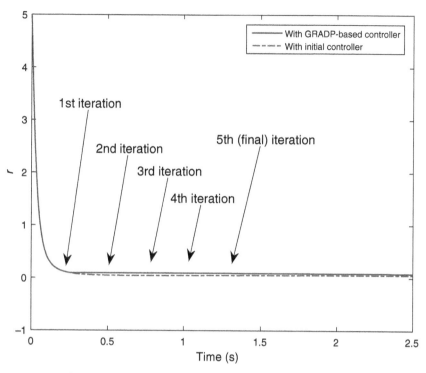

FIGURE 5.5 Simulation of the jet engine: trajectories of r.

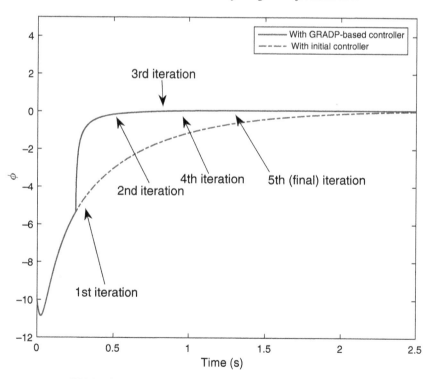

FIGURE 5.6 Simulation of the jet engine: trajectories of ϕ.

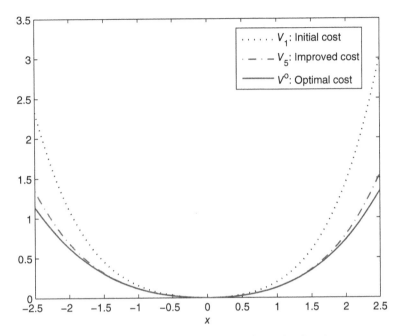

FIGURE 5.7 Simulation of the jet engine: value functions.

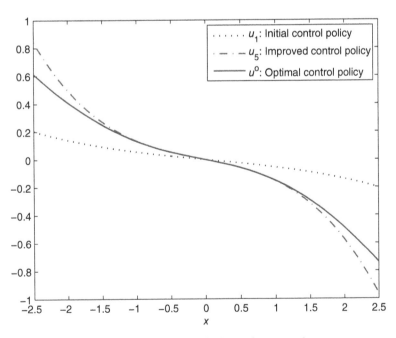

FIGURE 5.8 Simulation of the jet engine: control curve.

5.4 NOTES

Classical dynamic programming (DP) requires the precise knowledge of the system dynamics, the so-called *curse of modeling* in [2]. This strong requirement significantly limits the applicability of classic DP for real-world applications, which naturally involve model uncertainties. Robust dynamic programming (RDP) approaches have been proposed to robustify dynamic programming algorithms such that they can converge to a suboptimal solution, in the presence of static uncertainties (see, e.g., [3, 5], and [17]). While RDP enhances robustness of DP algorithms with respect to model uncertainties at the expense of better system performance, ADP approaches introduced in Chapters 2–4 actively adjust the control policy by using real-time measurements to achieve optimal performance of the closed-loop system.

In the past literature of both ADP and RDP, it is commonly assumed that the system order is known and the state variables are fully available. In practice, the system order may be unknown due to the presence of dynamic uncertainties (or unmodeled dynamics) [10], which are motivated by engineering applications in situations where the exact mathematical model of a physical system is not easy to be obtained. Of course, dynamic uncertainties also make sense for the mathematical modeling in other branches of science such as biology and economics. This problem, often formulated in the context of robust control theory, cannot be viewed as a special case of output feedback control. In addition, the ADP methods developed in the past literature may fail to guarantee not only optimality, but also the stability of the closed-loop system when dynamic uncertainty occurs. In fact, the performance of ADP learning may deteriorate with incomplete data or partial-state measurements [24].

The RADP methodologies introduced in this chapter can be viewed as natural extensions of ADP to dynamically perturbed uncertain systems. The RADP framework decomposes the uncertain environment into two parts: the *reduced-order system* (ideal environment) with known system order and fully accessible state, and the *dynamic uncertainties*, with unknown system order and dynamics, interacting with the ideal environment. The presence of dynamic uncertainty gives rise to interconnected systems for which the controller design and robustness analysis become technically challenging. In order to perform stability analysis of the interconnected systems, we adopt the notions of input-to-state stability (ISS), input-to-output stability (IOS), and strong unboundedness observability (SUO), as introduced in the nonlinear control community (see, for instance, [4, 11], and [21]). The robust stability and the suboptimality properties for the overall interconnected systems are realized by means of Lyapunov and small-gain techniques [11]. RADP is different from the online learning methods based on the theory of zero-sum games [1], which cannot always guarantee an arbitrarily small gain from the disturbance input to the output.

In this chapter, the RADP methods for partially linear composite systems and nonlinear systems are developed based on [7] and [8], respectively. RADP for nonlinear systems with unmatched disturbance and more detailed implementation issues can be found in [8]. In the next two chapters, RADP methodology will be applied to large-scale systems and stochastic systems, with applications to multi-machine power systems and human sensorimotor control systems.

REFERENCES

[1] A. Al-Tamimi, F. L. Lewis, and M. Abu-Khalaf. Model-free Q-learning designs for linear discrete-time zero-sum games with application to H_∞ control. *Automatica*, 43(3):473–481, 2007.

[2] D. P. Bertsekas and J. N. Tsitsiklis. *Neuro-Dynamic Programming*. Athena Scientific, Nashua, NH, 1996.

[3] M. Diehl. Robust dynamic programming for min-max model predictive control of constrained uncertain systems. *IEEE Transactions on Automatic Control*, 49(12):2253–2257, 2004.

[4] A. Isidori. *Nonlinear Control Systems*, Vol. 2. Springer, 1999.

[5] G. N. Iyengar. Robust dynamic programming. *Mathematics of Operations Research*, 30(2):257–280, 2005.

[6] Y. Jiang and Z. P. Jiang. Robust adaptive dynamic programming. In: D. Liu and F. Lewis, editors, *Reinforcement Learning and Adaptive Dynamic Programming for Feedback Control*, Chapter 13, pp. 281–302. John Wiley & Sons, 2012.

[7] Y. Jiang and Z. P. Jiang. Robust adaptive dynamic programming with an application to power systems. *IEEE Transactions on Neural Networks and Learning Systems*, 24(7):1150–1156, 2013.

[8] Y. Jiang and Z. P. Jiang. Robust adaptive dynamic programming and feedback stabilization of nonlinear systems. *IEEE Transactions on Neural Networks and Learning Systems*, 25(5):882–893, 2014.

[9] Z. P. Jiang, I. M. Mareels, and Y. Wang. A Lyapunov formulation of the nonlinear small-gain theorem for interconnected ISS systems. *Automatica*, 32(8):1211–1215, 1996.

[10] Z. P. Jiang and L. Praly. Design of robust adaptive controllers for nonlinear systems with dynamic uncertainties. *Automatica*, 34(7):825–840, 1998.

[11] Z. P. Jiang, A. R. Teel, and L. Praly. Small-gain theorem for ISS systems and applications. *Mathematics of Control, Signals and Systems*, 7(2):95–120, 1994.

[12] H. K. Khalil. *Nonlinear Systems*, 3rd ed. Prentice Hall, Upper Saddle River, NJ, 2002.

[13] M. Krstic, D. Fontaine, P. V. Kokotovic, and J. D. Paduano. Useful nonlinearities and global stabilization of bifurcations in a model of jet engine surge and stall. *IEEE Transactions on Automatic Control*, 43(12):1739–1745, 1998.

[14] M. Krstic, I. Kanellakopoulos, and P. V. Kokotovic. *Nonlinear and Adaptive Control Design*. John Wiley & Sons, New York, 1995.

[15] P. Kundur, N. J. Balu, and M. G. Lauby. *Power System Stability and Control*, Vol. 7. McGraw-Hill, New York, 1994.

[16] F. Lewis, D. Vrabie, and V. Syrmos. *Optimal Control*, 3rd ed. John Wiley & Sons, Inc., Hoboken, NJ, 2012.

[17] A. Madievsky and J. Moore. On robust dynamic programming. In: Proceedings of the 1996 IFAC World Congress, pp. 273–278, San Francisco, 1996.

[18] F. Moore and E. Greitzer. A theory of post-stall transients in axial compression systems: Part I — Development of equations. *Journal of engineering for gas turbines and power*, 108(1):68–76, 1986.

[19] A. Saberi, P. Kokotovic, and H. Summann. Global stabilization of partially linear composite systems. *SIAM*, 2(6):1491–1503, 1990.

[20] E. D. Sontag. Further facts about input to state stabilization. *IEEE Transactions on Automatic Control*, 35(4):473–476, 1990.

[21] E. D. Sontag. Input to state stability: Basic concepts and results. In: *Nonlinear and Optimal Control Theory*, pp. 163–220. Springer, 2008.

[22] E. D. Sontag and Y. Wang. On characterizations of the input-to-state stability property. *Systems & Control Letters*, 24(5):351–359, 1995.

[23] D. Vrabie and F. Lewis. Adaptive dynamic programming algorithm for finding online the equilibrium solution of the two-player zero-sum differential game. In: Proceedings of the 2010 IEEE International Joint Conference on Neural Networks (IJCNN), pp. 1–8, Barcelona, Spain, 2010.

[24] P. J. Werbos. Intelligence in the brain: A theory of how it works and how to build it. *Neural Networks*, 22(3):200–212, 2009.

CHAPTER 6

ROBUST ADAPTIVE DYNAMIC PROGRAMMING FOR LARGE-SCALE SYSTEMS

This chapter extends the RADP theory introduced in Chapter 5 for the decentralized optimal control of a generalized class of large-scale systems. See Figure 6.1 for the block diagram of such a setting. The controller design of each subsystem only utilizes local state variables, without knowing the system dynamics. By integrating a simple version of the cyclic-small-gain theorem, asymptotic stability can be achieved by assigning appropriate weighting matrices for each subsystem. As a by-product, certain suboptimality properties can be obtained.

6.1 STABILITY AND OPTIMALITY FOR LARGE-SCALE SYSTEMS

This section describes the class of large-scale uncertain systems to be studied. Then, an RADP-based decentralized optimal controller design scheme is presented. It will also be shown that the closed-loop interconnected system enjoys some suboptimality properties.

Robust Adaptive Dynamic Programming, First Edition. Yu Jiang and Zhong-Ping Jiang.
© 2017 by The Institute of Electrical and Electronics Engineers, Inc. Published 2017 by John Wiley & Sons, Inc.

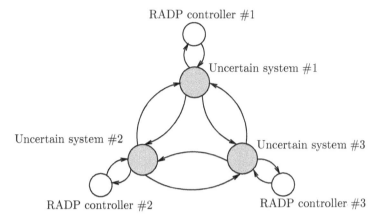

FIGURE 6.1   RADP-based online learning control for uncertain large-scale systems.

6.1.1 Description of Large-Scale Systems

Consider the complex large-scale system, of which the i-subsystem ($1 \leq i \leq N$) is described by

$$\dot{x}_i = A_i x_i + B_i [u_i + \Psi_i(y)], \tag{6.1}$$

$$y_i = C_i x_i, \ 1 \leq i \leq N \tag{6.2}$$

where $x_i \in \mathbb{R}^{n_i}$, $y_i \in \mathbb{R}^{q_i}$, and $u_i \in \mathbb{R}^{m_i}$ are the state, the output, and the control input for the ith subsystem; $y = \left[y_1^T, y_2^T, \dots, y_N^T \right]^T$; and $A_i \in \mathbb{R}^{n_i \times n_i}$, $B_i \in \mathbb{R}^{n_i \times m_i}$ are the unknown system matrices. $\Psi_i(\cdot) : \mathbb{R}^q \to \mathbb{R}^{m_i}$ are unknown interconnections satisfying $|\Psi_i(y)| \leq d_i |y|$ for all $y \in \mathbb{R}^q$, with $d_i > 0$, $\sum_{i=1}^N n_i = n$, $\sum_{i=1}^N q_i = q$, and $\sum_{i=1}^N m_i = m$. It is also assumed that (A_i, B_i) is a stabilizable pair, that is, there exists a constant matrix K_i such that $A_i - B_i K_i$ is a stable matrix.

Notice that the decoupled system can be written in a compact form

$$\dot{x} = A_D x + B_D u \tag{6.3}$$

where

$$x = \left[x_1^T, x_2^T, \dots, x_N^T \right]^T \in \mathbb{R}^n,$$

$$u = \left[u_1^T, u_2^T, \dots, u_N^T \right]^T \in \mathbb{R}^m,$$

$$A_D = \text{block diag}(A_1, A_2, \dots, A_N) \in \mathbb{R}^{n \times n},$$

$$B_D = \text{block diag}(B_1, B_2, \dots, B_N) \in \mathbb{R}^{n \times m}.$$

For system (6.3), we define the following quadratic cost

$$J_D = \int_0^\infty \left(x^T Q_D x + u^T R_D u \right) d\tau \tag{6.4}$$

where

$$Q_D = \text{block diag}(Q_1, Q_2, \ldots, Q_N) \in \mathbb{R}^{n \times n}, \tag{6.5}$$

$$R_D = \text{block diag}(R_1, R_2, \ldots, R_N) \in \mathbb{R}^{m \times m}, \tag{6.6}$$

$Q_i \in \mathbb{R}^{n_i \times n_i}$, and $R_i \in \mathbb{R}^{m_i \times m_i}$, with $Q_i = Q_i^T \geq 0, R_i = R_i^T > 0$, and $(A_i, Q_i^{1/2})$ observable, for all $1 \leq i \leq N$.

By linear optimal control theory [16], a minimum cost $J_D^\odot$ described in (6.4) can be obtained for the decoupled system by employing the following decentralized control policy

$$u_D^\odot = -K_D x \tag{6.7}$$

where $K_D = \text{block diag}(K_1, K_2, \ldots, K_N)$ is given by

$$K_D = R_D^{-1} B_D^T P_D \tag{6.8}$$

and $P_D = \text{block diag}(P_1, P_2, \ldots, P_N)$ is the unique symmetric positive definite solution of the algebraic Riccati equation (ARE)

$$A_D^T P_D + P_D A_D - P_D B_D R_D^{-1} B_D^T P_D + Q_D = 0 \tag{6.9}$$

6.1.2 Decentralized Stabilization

Now, we analyze the stability of the closed-loop system comprised of (6.1), (6.2) and the decentralized controller (6.7). We show that by selecting appropriate weighting matrices Q_D and R_D, global asymptotic stability can be achieved for the large-scale closed-loop system.

To begin with, let us introduce two lemmas.

Lemma 6.1.1   *For any $\gamma_i > 0$ and $\epsilon_i > 0$, let $u^\odot$ be the decentralized control policy obtained from (6.7) to (6.9) with $Q_i \geq (\gamma_i^{-1} + 1)C_i^T C_i + \gamma_i^{-1} \epsilon_i I_{n_i}$ and $R_i^{-1} \geq d_i^2 I_{m_i}$. Then, along the solutions of the closed-loop system consisting of (6.1), (6.2), and (6.7), we have*

$$\frac{d}{dt} \left(\gamma_i x_i^T P_i x_i \right) \leq -|y_i|^2 - \epsilon_i |x_i|^2 + \gamma_i \sum_{j=1, j \neq i}^N |y_j|^2 \tag{6.10}$$

Proof: Along the solutions of the closed-loop system, we have

$$
\begin{aligned}
\frac{d}{dt}\left(x_i^T P_i x_i\right) &= x_i^T P_i \left[A_i x_i + B_i u_i + B_i \Psi_i(y)\right] \\
&\quad + \left[A_i x_i + B_i u_i + B_i \Psi_i(y)\right]^T P_i x_i \\
&= x_i^T P_i \left[A_i x_i - B_i K_i x_i + B_i \Psi_i(y)\right] \\
&\quad + \left[A_i x_i - B_i K_i x_i + B_i \Psi_i(y)\right]^T P_i x_i \\
&= x_i^T \left[P_i(A_i - B_i K_i) + (A_i - B_i K_i)^T P_i\right] x_i \\
&\quad + x_i^T P_i B_i \Psi_i(y) + \Psi_i^T(y) B_i^T P_i x_i \\
&= -x_i^T Q_i x_i - x_i^T P_i B_i R_i^{-1} B_i^T P_i x_i \\
&\quad + x_i^T P_i B_i \Psi_i(y) + \Psi_i^T(y) B_i^T P_i x_i \\
&\leq -x_i^T Q_i x_i - d_i^2 x_i^T P_i B_i B_i^T P_i x_i \\
&\quad + x_i^T P_i B_i \Psi_i(y) + \Psi_i^T(y) B_i^T P_i x_i \\
&= -x_i^T Q_i x_i + d_i^{-2} \Psi_i^T(y) \Psi_i(y) \\
&\quad - d_i^2 x_i^T P_i B_i B_i^T P_i x_i + x_i^T P_i B_i \Psi_i(y) \\
&\quad + \Psi_i^T(y) B_i^T P_i x_i - d_i^{-2} \Psi_i^T(y) \Psi_i(y) \\
&= -x_i^T Q_i x_i + d_i^{-2} \Psi_i^T(y) \Psi_i(y) \\
&\quad - \left[d_i B_i^T P_i x_i - d_i^{-1} \Psi_i(y)\right]^T \left[d_i B_i^T P_i x_i - d_i^{-1} \Psi_i(y)\right] \\
&\leq -x_i^T Q_i x_i + d_i^{-2} \Psi_i^T(y) \Psi_i(y) \\
&\leq -x_i^T \left[\left(\gamma_i^{-1} + 1\right) C_i^T C_i + \gamma_i^{-1} \epsilon_i I_{n_i}\right] x_i + d_i^{-2} \Psi_i^T(y) \Psi_i(y) \\
&\leq -\left(\gamma_i^{-1} + 1\right) |y_i|^2 - \gamma_i^{-1} \epsilon_i |x_i|^2 + |y|^2 \\
&\leq -\gamma_i^{-1} |y_i|^2 - \gamma_i^{-1} \epsilon_i |x_i|^2 + \sum_{j=1, j \neq i}^{N} |y_j|^2
\end{aligned}
$$

Therefore,

$$
\frac{d}{dt}\left(\gamma_i x_i^T P_i x_i\right) \leq -|y_i|^2 - \epsilon_i |x_i|^2 + \gamma_i \sum_{j=1, j \neq i}^{N} |y_j|^2
$$

The proof is complete. ∎

Lemma 6.1.2   *Under the conditions of Lemma 6.1.1, suppose the following cyclic-small-gain condition holds*

$$\sum_{j=1}^{N-1} j \sum_{1 \le i_1 < i_2 < \cdots < i_{j+1} \le j+1} \gamma_{i_1} \gamma_{i_2} \cdots \gamma_{i_{j+1}} < 1 \qquad (6.11)$$

Then, there exist constants $c_i > 0$ for all $1 \le i \le N$, such that along the solutions of the closed-loop system (6.1), (6.2), and (6.7), we have

$$\frac{d}{dt}\left(\sum_{i=1}^{N} c_i \gamma_i x_i^T P_i x_i \right) \le -|y|^2 - \sum_{i=1}^{N} c_i \gamma_i \epsilon_i |x_i|^2 \qquad (6.12)$$

Proof: To begin with, let us consider the following linear equations

$$\begin{bmatrix} -1 & \gamma_2 & \gamma_3 & \cdots & \gamma_N \\ \gamma_1 & -1 & \gamma_3 & \cdots & \gamma_N \\ \gamma_1 & \gamma_2 & -1 & \ddots & \gamma_N \\ \vdots & \vdots & \ddots & \ddots & \vdots \\ \gamma_1 & \gamma_2 & \gamma_3 & \cdots & -1 \end{bmatrix} \begin{bmatrix} c_1 \\ c_2 \\ c_3 \\ \vdots \\ c_N \end{bmatrix} = \begin{bmatrix} -1 \\ -1 \\ -1 \\ \vdots \\ -1 \end{bmatrix} \qquad (6.13)$$

First, we show that, if the cyclic-small-gain condition (6.11) holds, the equation (6.13) can be solved as

$$c_i = \frac{\prod_{j=1, j \ne i}^{N}(\gamma_j + 1)}{1 - \sum_{j=1}^{N-1} j \sum_{1 \le i_1 < i_2 < \cdots < i_{j+1} \le j+1} \gamma_{i_1} \gamma_{i_2} \cdots \gamma_{i_{j+1}}} > 0 \qquad (6.14)$$

Indeed, it can be proved by mathematical induction: (1) if $N = 2$, (6.13) is reduced to

$$\begin{bmatrix} -1 & \gamma_2 \\ \gamma_1 & -1 \end{bmatrix} \begin{bmatrix} c_1 \\ c_2 \end{bmatrix} = \begin{bmatrix} -1 \\ -1 \end{bmatrix} \qquad (6.15)$$

and the solution is $c_1 = \dfrac{1 + \gamma_2}{1 - \gamma_1 \gamma_2}$, $c_2 = \dfrac{1 + \gamma_1}{1 - \gamma_1 \gamma_2}$. Notice that the solution is unique, because the cyclic-small-gain condition (6.11) guarantees that the determinant of the coefficient matrix is non-zero.

(2) Suppose (6.14) is the solution of (6.13) with $N = N' - 1$, we show that it is also valid for $N = N'$. Then, from the first row of (6.13), we have

$$
-\prod_{j=1,j\neq i}^{N'-1} (\gamma_j + 1) + \sum_{i=2}^{N'-1} \prod_{j=1,j\neq i}^{N'-1} (\gamma_j + 1)\gamma_i
$$

$$
= \sum_{j=1}^{N'-2} j \sum_{1\leq i_1 < i_2 < \cdots < i_{j+1} \leq j+1} \gamma_{i_1} \gamma_{i_2} \cdots \gamma_{i_{j+1}} - 1 \tag{6.16}
$$

Now,

$$
-\prod_{j=1,j\neq i}^{N'} (\gamma_j + 1) + \sum_{i=2}^{N'} \prod_{j=1,j\neq i}^{N'} (\gamma_j + 1)\gamma_i
$$

$$
= -\prod_{j=1,j\neq i}^{N'-1} (\gamma_j + 1)(\gamma_{N'} + 1) + \sum_{i=2}^{N'-1} \prod_{j=1,j\neq i}^{N'-1} (\gamma_j + 1)(\gamma_{N'} + 1)\gamma_i
$$

$$
+ \prod_{j=1}^{N'-1} (\gamma_j + 1)\gamma_{N'}
$$

$$
= \left[-\prod_{j=1,j\neq i}^{N'-1} (\gamma_j + 1) + \sum_{i=2}^{N'-1} \prod_{j=1,j\neq i}^{N'-1} (\gamma_j + 1)\gamma_i \right] (\gamma_{N'} + 1)
$$

$$
+ \prod_{j=1}^{N'-1} (\gamma_j + 1)\gamma_{N'}
$$

$$
= \left[\sum_{j=1}^{N'-2} j \sum_{1\leq i_1 < i_2 < \cdots < i_{j+1} \leq j+1} \gamma_{i_1} \gamma_{i_2} \cdots \gamma_{i_{j+1}} - 1 \right] (\gamma_{N'} + 1)
$$

$$
+ \prod_{j=1}^{N'-1} (\gamma_j + 1)\gamma_{N'}
$$

$$
= \sum_{j=1}^{N'-1} j \sum_{2\leq i_1 < i_2 < \cdots < i_{j+1} \leq j+1} \gamma_{i_1} \gamma_{i_2} \cdots \gamma_{i_{j+1}}
$$

$$
+ \sum_{j=1}^{N'-2} j \sum_{1\leq i_1 < i_2 < \cdots < i_{j+1} \leq j+1} \gamma_{i_1} \gamma_{i_2} \cdots \gamma_{i_{j+1}}
$$

$$+ \prod_{j=1}^{N'-1} (\gamma_j + 1)\gamma_{N'} - \gamma_{N'} - 1$$

$$= \sum_{j=1}^{N'-1} j \sum_{1 \le i_1 < i_2 < \cdots < i_{j+1} \le j+1} \gamma_{i_1} \gamma_{i_2} \cdots \gamma_{i_{j+1}} - 1$$

This implies, with $N = N'$, the first row of (6.13) is valid with the solution (6.14). Same derivations can be applied to the remaining rows.

Together with Lemma 2.1 we obtain

$$\frac{d}{dt} \left(\sum_{i=1}^{N} c_i \gamma_i x_i^T P_i x_i \right)$$

$$\le - \sum_{i=1}^{N} c_i \gamma_i \epsilon_i |x_i|^2 + \sum_{i=1}^{N} c_i \left(-|y_i|^2 + \gamma_i \sum_{j=1, j \ne i}^{N} |y_j|^2 \right)$$

$$\le - \sum_{i=1}^{N} c_i \gamma_i \epsilon_i |x_i|^2 + \begin{bmatrix} |y_1|^2 \\ |y_2|^2 \\ |y_3|^2 \\ \cdots \\ |y_N|^2 \end{bmatrix}^T \begin{bmatrix} -1 & \gamma_2 & \gamma_3 & \cdots & \gamma_N \\ \gamma_1 & -1 & \gamma_3 & \cdots & \gamma_N \\ \gamma_1 & \gamma_2 & -1 & \ddots & \gamma_N \\ \vdots & \vdots & \ddots & \ddots & \vdots \\ \gamma_1 & \gamma_2 & \gamma_3 & \cdots & -1 \end{bmatrix} \begin{bmatrix} c_1 \\ c_2 \\ c_3 \\ \vdots \\ c_N \end{bmatrix}$$

$$= - \sum_{i=1}^{N} c_i \gamma_i \epsilon_i |x_i|^2 - |y|^2$$

The proof is complete. ∎

In summary, we obtain the following theorem.

Theorem 6.1.3   *The overall closed-loop system (6.1), (6.2), and (6.7) is globally asymptotically stable (GAS) if the cyclic-small-gain condition (6.11) holds.*

Proof: Define the Lyapunov function candidate

$$V_N = \sum_{i=1}^{N} c_i \gamma_i x_i^T P_i x_i \tag{6.17}$$

By Lemma 6.1.2, along the solutions of (6.1) and (6.2), it follows

$$\dot{V}_N \leq -\sum_{i=1}^{N} c_i \gamma_i \epsilon_i |x_i|^2 - |y|^2 \tag{6.18}$$

Hence, the closed-loop system is GAS at the origin.   ∎

Remark 6.1.4   *It is of interest to note that a more generalized cyclic-small-gain condition based on the notion of input-to-output stability [9, 31] can be found in [10, 11], and [19].*

6.1.3 Suboptimality Analysis

Suppose $\Psi_i(\cdot)$ is differentiable at the origin for all $1 \leq i \leq N$, system (6.1) and (6.2) can be linearized around the origin as

$$\dot{x} = A_D x + B_D u + A_C x, \ y = C_D x \tag{6.19}$$

As pointed out in [28], under the decentralized control policy (6.7), the cost (6.4) yields a finite cost $J_D^{\oplus}$ for the coupled system (6.19), which may differ from $J_D^{\odot}$. In order to study the relationship between $J_D^{\oplus}$ and $J_D^{\odot}$, define

$$M_D = \text{block diag}(\sigma_1 I_{n_1}, \sigma_2 I_{n_2}, \cdots, \sigma_N I_{n_N}) \in \mathbb{R}^{n \times n} \tag{6.20}$$

where $\sigma_i > 0$ with $1 \leq i \leq N$.

To quantify the suboptimality of the closed-loop system composed of (6.19) and (6.7), we recall the following concept and theorem from [28].

Definition 6.1.5 ([28])   *The decentralized control law (6.7) is said to be suboptimal with degree σ for system (6.19), if there exists a positive number σ such that*

$$J_D^{\oplus} \leq \sigma^{-1} J_D^{\odot} \tag{6.21}$$

Theorem 6.1.6 ([28])   *Suppose there exists a matrix M_D as defined in (6.20) such that the matrix*

$$F(M_D) = A_C^T M_D^{-1} P_D + M_D^{-1} P_D A_C + (I - M_D^{-1})(Q_D + K_D^T R_D K_D) \tag{6.22}$$

satisfies $F(M_D) \leq 0$. Then, the control $u_D^{\odot}$ is suboptimal for (6.19) with degree

$$\sigma = \min_{1 \leq i \leq N} \{\sigma_i\} \tag{6.23}$$

The following theorem summarizes the suboptimality of the controller (6.7) under the cyclic-small-gain condition (6.11).

Theorem 6.1.7   *The decentralized controller $u_D^\odot$ is suboptimal for system (6.1) and (6.2) with degree*

$$\sigma = \min_{1 \leq i \leq N} \left\{ \frac{1}{c_i \gamma_i} \min_{1 \leq i \leq N} \left(\frac{c_i \gamma_i}{\gamma_i \epsilon_i^{-1} \lambda_M + 1}, 1 \right) \right\} \tag{6.24}$$

if the condition (6.11) holds.

Proof: Let $\sigma_i^{-1} = \alpha c_i \gamma_i$ with $\alpha > \dfrac{1}{\min\limits_{1 \leq i \leq N} (c_i \gamma_i)}$ and $\alpha \geq 1$.

Then, by (6.22) we obtain

$$\frac{d}{dt} \left(x^T M_D^{-1} P_D x \right)$$
$$= x^T \left(M_D^{-1} P_D A_D + A_D M_D^{-1} P_D + M_D^{-1} P_D A_C + A_C^T M_D^{-1} P_D \right) x$$
$$= x^T \left(-M_D^{-1} Q_D - M_D^{-1} K_D^T R_D K_D + M_D^{-1} P_D A_C + A_C^T M_D^{-1} P_D \right) x$$
$$= x^T \left[F(M_D) - Q_D - K_D^T R_D K_D \right] x$$

Therefore, by Lemma 6.1.2, it follows that

$$x^T F(M_D) x = \frac{d}{dt} \left(x^T M_D^{-1} P_D x \right) + x^T \left(Q_D + K_D^T R_D K_D \right) x$$

$$\leq \sum_{i=1}^{N} x_i^T \left[-\sigma_i^{-1} (Q_i - C_i^T C_i) - \alpha |y_i|^2 + Q_i + K_i^T R_i K_i \right] x_i$$

$$= -\sum_{i=1}^{N} x_i^T \left[\sigma_i^{-1} (Q_i - C_i^T C_i) + \alpha C_i^T C_i - Q_i - K_i^T R_i K_i \right] x_i$$

$$= -\sum_{i=1}^{N} x_i^T \left[(\sigma_i^{-1} - 1) (Q_i - C_i^T C_i) - K_i^T R_i K_i \right] x_i$$

$$\leq -\sum_{i=1}^{N} \left[(\alpha c_i \gamma_i - 1) \frac{\epsilon_i}{\gamma_i} - \lambda_M \right] |x_i|^2$$

where λ_M denotes the maximal eigenvalue of $K_i^T R_i K_i$.

Notice that $F(M_D) \leq 0$, if we set

$$\alpha = \max_{1 \leq i \leq N} \left(\frac{\gamma_i \epsilon_i^{-1} \lambda_M + 1}{c_i \gamma_i}, 1 \right) \tag{6.25}$$

Therefore, we obtain

$$\sigma = \min_{1 \leq i \leq N} \left\{ \frac{1}{c_i \gamma_i} \min_{1 \leq i \leq N} \left(\frac{c_i \gamma_i}{\gamma_i \epsilon_i^{-1} \lambda_M + 1}, 1 \right) \right\} \tag{6.26}$$

The proof is complete by Theorem 6.1.6. ■

6.2 RADP FOR LARGE-SCALE SYSTEMS

Consider the following ARE

$$A_i^T P_i + P_i A_i + Q_i - P_i B_i R_i^{-1} B_i^T P_i = 0, 1 \leq i \leq N \tag{6.27}$$

As shown in [13], given $K_i^{(0)}$ such that $A_i - B_i K_i^{(0)}$ is Hurwitz, the sequences $\{P_i^{(k)}\}$ and $\{K_i^{(k)}\}$ uniquely determined by

$$0 = (A_i^{(k)})^T P_i^{(k)} + P_i A_i^{(k)} + Q_i^{(k)}, \tag{6.28}$$

$$K_i^{(k+1)} = R_i^{-1} B_i^T P_i^{(k)} \tag{6.29}$$

with $A_i^{(k)} = A_i - B_i K_i^{(k)}$, and $Q_i^{(k)} = Q_i + (K_i^{(k)})^T R_i K_i^{(k)}$, enjoy the following properties:

(1) $\lim_{k \to \infty} P_i^{(k)} = P_i$,

(2) $\lim_{k \to \infty} K_i^{(k)} = K_i = R_i^{-1} B_i^T P_i$, and

(3) $A_i^{(k)}$ is Hurwitz for all $k \in \mathbb{Z}_+$.

For the ith subsystem, along the solutions of (6.1) and (6.2), it follows that

$$x_i^T P_i^{(k)} x_i \Big|_t^{t+\delta t} = 2 \int_t^{t+\delta t} (\hat{u}_i + K_i^{(k)} x_i)^T R_i K_i^{(k+1)} x_i d\tau$$

$$- \int_t^{t+\delta t} x_i^T Q_i^{(k)} x_i d\tau \tag{6.30}$$

where $\hat{u}_i = u_i + \Psi_i(y)$.

For any sufficiently large integer $l_i \geq 0$, define $\delta_{xx}^i \in \mathbb{R}^{l_i \times n_i^2}$, $I_{xx}^i \in \mathbb{R}^{l_i \times n_i^2}$, and $I_{xu}^i \in \mathbb{R}^{l_i \times m_i n_i}$ as follows:

$$\delta_{xx}^i = \left[x_i \otimes x_i |_{t_{0,i}}^{t_{1,i}} \quad x_i \otimes x_i |_{t_{1,i}}^{t_{2,i}} \quad \cdots \quad x_i \otimes x_i |_{t_{l_i-1,i}}^{t_{l_i,i}} \right]^T,$$

$$I_{xx}^i = \left[\int_{t_{0,i}}^{t_{1,i}} x_i \otimes x_i d\tau \quad \int_{t_{1,i}}^{t_{2,i}} x_i \otimes x_i d\tau \quad \cdots \quad \int_{t_{l_i-1,i}}^{t_{l_i,i}} x_i \otimes x_i d\tau \right]^T,$$

$$I_{xu}^i = \left[\int_{t_{0,i}}^{t_{1,i}} x_i \otimes \hat{u}_i d\tau \quad \int_{t_{1,i}}^{t_{2,i}} x_i \otimes \hat{u}_i d\tau \quad \cdots \quad \int_{t_{l_i-1,i}}^{t_{l_i,i}} x_i \otimes \hat{u}_i d\tau \right]^T$$

where $0 \leq t_{0,i} < t_{1,i} < \cdots < t_{l_i,i}$, for $i = 1, 2, \ldots, N$. Then, (6.30) implies the following matrix form of linear equations

$$\Theta_i^{(k)} \left[\begin{array}{c} \text{vec}(P_i^{(k)}) \\ \text{vec}(K_i^{(k+1)}) \end{array} \right] = \Xi_i^{(k)} \tag{6.31}$$

where the matrices $\Theta_i^{(k)} \in \mathbb{R}^{l_i \times n_i(n_i + m_i)}$ and $\Xi_i^{(k)} \in \mathbb{R}^{l_i}$ are defined as

$$\Theta_i^{(k)} = \left[\delta_{xx}^i \quad -2I_{xx}^i (I_{n_i} \otimes (K_i^{(k)})^T R_i) - 2I_{xu}^i (I_{n_i} \otimes R_i) \right],$$

$$\Xi_i^{(k)} = -I_{xx}^i \text{vec}(Q_i^{(k)}).$$

Clearly, if (6.31) has a unique solution satisfying $P_i^{(k)} = P_i^{(k)T}$, we are able to replace (6.28) and (6.29) by (6.31). In this way, the knowledge of both A_i and B_i is no longer needed.

Similar as in Assumption 2.3.8, a rank condition is assumed as follows.

Assumption 6.2.1   *The following rank condition holds*

$$\text{rank}\left([I_{xx}^i, I_{xu}^i] \right) = \frac{n_i(n_i + 1)}{2} + n_i m_i. \tag{6.32}$$

Next, we give the following robust ADP algorithm for practical online implementation. Notice that the algorithm can be implemented to each subsystem in parallel without affecting each other. The learning system implemented for each subsystem only needs to use the state and input information of the subsystem.

Algorithm 6.2.2   *Robust ADP algorithm for large-scale systems*

(1) *Initialization:*
   Select appropriate matrices Q_i and R_i such that the condition (6.11) is satisfied.
   Let $k \leftarrow 0$.

(2) *Online data collection:*
   For the ith subsystem, employ $u_i = -K_i^{(0)} x_i + e_i$, with e_i the exploration noise, as the input. Compute δ_{xx}^i, I_{xx}^i and I_{xu}^i until Assumption 6.2.1 is satisfied.

(3) *Policy iteration and policy improvement:*
   Solve $P_i^{(k)}$ and $K_i^{(k+1)}$ from (6.31). Then, repeat Step (3) with $k \leftarrow k + 1$, until

$$|P_i^{(k)} - P_i^{(k-1)}| \leq \epsilon \tag{6.33}$$

   for all $k \geq 1$, where the constant $\epsilon > 0$ can be any predefined small threshold.

(4) *Exploitation:*
   Use $u_i = -K_i^{(k)} x_i$ as the approximated optimal control policy to the ith subsystem.

Remark 6.2.3   *Note that the convergence property of Algorithm 6.2.2 is assured under Assumption 6.2.1. It can be proved following the same logic as in the proof of Lemma 2.3.9.*

6.3 EXTENSION FOR SYSTEMS WITH UNMATCHED DYNAMIC UNCERTAINTIES

Consider interconnected systems described by the following equations with $i = 1, 2, \ldots, N$

$$\dot{x}_i = A_i x_i + B_i(z_i + \Psi_i(y)), \tag{6.34}$$

$$\dot{z}_i = E_i x_i + F_i z_i + G_i(u_i + \Phi_i(y)), \tag{6.35}$$

$$y_i = C_i x_i, \tag{6.36}$$

where $[x_i^T, z_i^T]^T \in \mathbb{R}^{n_i} \times \mathbb{R}^{m_i}$, $u_i \in \mathbb{R}^{m_i}$, and $y_i \in \mathbb{R}^{q_i}$ are the state, input, and output of the ith subsystem, respectively; $y = [y_1^T, y_2^T, \ldots, y_N^T]^T$; $N > 0$ is the number of subsystems; $A_i \in \mathbb{R}^{n_i \times n_i}$, $B_i \in \mathbb{R}^{n_i \times m_i}$, $E_i \in \mathbb{R}^{m_i \times n_i}$, $F_i \in \mathbb{R}^{m_i \times m_i}$, $G_i \in \mathbb{R}^{m_i \times m_i}$, and $C_i \in \mathbb{R}^{q_i \times n_i}$ are unknown matrices with (A_i, B_i) stabilizable and G_i nonsingular; $\Psi_i : \mathbb{R}^q \to \mathbb{R}^{m_i}$ and $\Phi_i : \mathbb{R}^q \to \mathbb{R}^{m_i}$, with $q = \sum_{i=1}^{N} q_i$, are interconnections in the large-scale system. Suppose the exact forms of functions Ψ_i and Φ_i are unknown, and we only have $|\Psi_i(y)| \leq \gamma_i |y|$, $|\Phi_i(y)| \leq \eta_i |y|$, where γ_i and η_i are known positive constants.

6.3.1 Robust Optimal Design

The robust optimal control policy for the decoupled large-scale system (6.34)–(6.36) is developed by implementing the control gain K_i with the backstepping technique [14].

To begin with, denote $w_i = z_i + K_i x_i$. By (6.34) and (6.35), we have

$$\dot{w}_i = \dot{z}_i + K_i \dot{x}_i$$
$$= \bar{F}_i w_i + \bar{E}_i x_i + G_i u_i + G_i \Phi_i(y) + K_i B_i \Psi_i(y) \tag{6.37}$$

where $\bar{F}_i = F_i + K_i B_i$, and $\bar{E}_i = E_i + K_i A_i - (F_i + K_i B_i)K_i$. Then, we can rewrite the system (6.34)–(6.37) in the following form

$$\dot{\zeta}_i = A_{D,i}\zeta_i + B_{D,i}u_i + A_{\Delta,i}\Delta_i(y),$$
$$y_i = C_{D,i}\zeta_i \tag{6.38}$$

where $\zeta_i = \begin{bmatrix} x_i^T \\ w_i^T \end{bmatrix}$, $\Delta_i(y) = \begin{bmatrix} \Psi_i^T(y) \\ \Phi_i^T(y) \end{bmatrix}$, $A_{D,i} = \begin{bmatrix} A_i - B_i K_i & B_i \\ \bar{E}_i & \bar{F}_i \end{bmatrix}$, $B_{D,i} = \begin{bmatrix} 0 \\ G_i \end{bmatrix}$, $A_{\Delta,i} = \begin{bmatrix} B_i & 0 \\ K_i B_i & G_i \end{bmatrix}$, $C_{D,i} = [C_i \quad 0]$.

In the following lemma, we give the stability analysis for the reduced-order system of (6.38) (i.e., system (6.38) with $\Delta_i \equiv 0$).

Lemma 6.3.1 *Consider system (6.38) with $\Delta_i \equiv 0$, $i = 1, 2, \ldots, N$. Choose $Q_i = Q_i^T \in \mathbb{R}^{n_i \times n_i}$, $R_i = R_i^T \in \mathbb{R}^{m_i \times m_i}$, $S_i = S_i^T \in \mathbb{R}^{m_i \times m_i}$, $W_i = W_i^T \in \mathbb{R}^{m_i \times m_i}$, $\delta_i \in \mathbb{R}$, and $\varepsilon_i \in \mathbb{R}$, such that $\varepsilon_i > 0$, $\delta_i > 0$,*

$$Q_i \geq (\varepsilon_i + \delta_i)I_{n_i}, \tag{6.39}$$
$$R_i > 0, \tag{6.40}$$
$$S_i^{-1} \geq \varepsilon_i^{-1} G_i^{-1} \bar{E}_i \bar{E}_i^T G_i^{-T}, \tag{6.41}$$
$$W_i \geq B_i^T P_i^* (Q_i - (\varepsilon_i + \delta_i)I_{n_i})^{-1} P_i^* B_i + \delta_i I_{m_i}, \tag{6.42}$$

and that $(\bar{F}_i, G_i)$ is stabilizable, $(A_i, Q_i^{1/2})$ and $(\bar{F}_i, W_i^{1/2})$ are observable. Then, system (6.38) is made GAS at the origin with the controller

$$u_i = u_i^*(w_i) = -L_i^* w_i \tag{6.43}$$

where $L_i^* = S_i^{-1} G_i^T M_i^*$, with $M_i^* = M_i^{*T} > 0$ the unique solution to the following ARE:

$$\bar{F}_i^T M_i + M_i \bar{F}_i - M_i G_i S_i^{-1} G_i^T M_i + W_i = 0 \tag{6.44}$$

Proof: Choose the Lyapunov function candidate as

$$V_i(x_i, w_i) = x_i^T P_i^* x_i + w_i^T M_i^* w_i \tag{6.45}$$

Taking the derivative of V_i along the solutions to (6.38), we have

$$
\begin{aligned}
\dot{V}_i &= -\zeta_i^T \bar{Q}_i \zeta_i - w_i^T M_i^* G_i S_i^{-1} G_i^T M_i^* w_i \\
&\leq \zeta_i^T \begin{bmatrix} -Q_i + \varepsilon_i I_{n_i} & P_i^* B_i \\ B_i^T P_i^* & -W_i \end{bmatrix} \zeta_i - x_i^T P_i^* B_i R_i^{-1} B_i^T P_i^* x_i \\
&\quad - w_i^T M_i^* \left(G_i S_i^{-1} G_i^T - \epsilon_i^{-1} \bar{E}_i \bar{E}_i^T \right) M_i^* w_i \\
&\leq -\delta_i \zeta_i^T \zeta_i,
\end{aligned}
\tag{6.46}
$$

where

$$
\bar{Q}_i = \begin{bmatrix} Q_i + P_i^* B_i R_i^{-1} B_i^T P_i^* & -P_i^* B_i - \bar{E}_i^T M_i^* \\ -B_i^T P_i^* - M_i^* \bar{E}_i & W_i \end{bmatrix}
\tag{6.47}
$$

By classical Lyapunov theory [12], the proof follows readily from (6.46). ∎

Since $w_i = z_i + K_i x_i$, it is easy to see from Lemma 6.3.1 that the system (6.34)–(6.36) with $\Delta_i \equiv 0$ is GAS at the origin.

6.3.2 Stability Analysis

Next, we give the stability analysis for the large-scale system (6.34)–(6.36) under the control policy u_i^* developed in Lemma 6.3.1.

Theorem 6.3.2   *Consider system (6.38) with $u_i = -L_i^* w_i$. For a given $h_i > 0$, choose $Q_i = Q_i^T \in \mathbb{R}^{n_i \times n_i}$, $R_i = R_i^T \in \mathbb{R}^{m_i \times m_i}$, $S_i = S_i^T \in \mathbb{R}^{m_i \times m_i}$, $W_i = W_i^T \in \mathbb{R}^{m_i \times m_i}$, $\delta_i \in \mathbb{R}$, $\varepsilon_i \in \mathbb{R}$, $\epsilon_i \in \mathbb{R}$, and $\alpha_i \in \mathbb{R}$, such that $\varepsilon_i > 0$, $\epsilon_i > 0$, $\delta_i > 0$, $0 < \alpha_i < 1$,*

$$
Q_i \geq \left(\epsilon_i h_i^{-1} + \varepsilon_i + \delta_i \right) I_{n_i} + \left(1 + h_i^{-1} \right) C_i^T C_i,
$$

$$
R_i^{-1} \geq 2\alpha_i^{-1} \gamma_i^2 I_{m_i},
$$

$$
S_i^{-1} \geq \epsilon_i^{-1} G_i^{-1} \bar{E}_i \bar{E}_i^T G_i^{-T} + 2\eta_i^2 I_{m_i}
$$

$$
\qquad + 2(1 - \alpha_i)^{-1} \gamma_i^2 G_i^{-1} K_i B_i B_i^T K_i^T G_i^{-T},
$$

$$
W_i \geq B_i^T P_i^* \left(Q_i - \left(\epsilon_i h_i^{-1} + \varepsilon_i + \delta_i \right) I_{n_i} \right.
$$

$$
\qquad \left. - \left(1 + h_i^{-1} \right) C_i^T C_i \right)^{-1} P_i^* B_i + \left(\epsilon_i h_i^{-1} + \delta_i \right) I_{m_i},
$$

and that $(\bar{F}_i, G_i)$ is stabilizable, $(A_i, Q_i^{1/2})$ and $(\bar{F}_i, W_i^{1/2})$ are observable. Then,

$$
\frac{d}{dt} \left(x_i^T P_i^* x_i + w_i^T M_i^* w_i \right) \leq -\epsilon_i h_i^{-1} |\zeta_i|^2 - h_i^{-1} |y_i|^2
$$

$$
\qquad + \sum_{j=1, j \neq i}^N |y_j|^2
\tag{6.48}
$$

Proof: Choose the Lyapunov function candidate defined in (6.45). Differentiating V_i along the solutions of (6.38), we have

$$
\begin{aligned}
\dot{V}_i \leq\ & \zeta_i^T \Xi_i \zeta_i - x_i^T P_i^* B_i R_i^{-1} B_i^T P_i^* x_i - \frac{\epsilon_i}{h_i} x_i^T x_i - \frac{\epsilon_i}{h_i} w_i^T w_i \\
& - \left(1 + \frac{1}{h_i}\right) y_i^T y_i - w_i^T M_i^* (G_i S_i^{-1} G_i^T - \epsilon_i^{-1} \bar{E}_i \bar{E}_i^T) M_i^* w_i \\
& + x_i^T P_i^* B_i \Psi_i(y) + \Psi_i^T(y) B_i^T P_i^* x_i \\
& + w_i^T M_i^* (G_i \Phi_i(y) + K_i B_i \Psi_i(y)) \\
& + (G_i \Phi_i(y) + K_i B_i \Psi_i(y))^T M_i^* w_i
\end{aligned}
\tag{6.49}
$$

where

$$
\Xi_i =
\begin{bmatrix}
\left(\frac{\epsilon_i}{h_i} + \epsilon_i\right) I_{n_i} + \left(1 + \frac{1}{h_i}\right) C_i^T C_i - Q_i & P_i^* B_i \\
B_i^T P_i^* & \frac{\epsilon_i}{h_i} I_{m_i} - W_i
\end{bmatrix}
\tag{6.50}
$$

By completing the squares, it follows successively that

$$
\begin{aligned}
& x_i^T P_i^* B_i \Psi_i(y) + \Psi_i^T(y) B_i^T P_i^* x_i \\
& \leq \frac{1}{2} \alpha_i \gamma_i^{-2} \Psi_i^T(y) \Psi_i(y) + 2\alpha_i^{-1} \gamma_i^2 x_i^T P_i^* B_i B_i^T P_i^* x_i, \\
& \times w_i^T M_i^* G_i \Phi_i(y) + \Phi_i^T(y) G_i^T M_i^* w_i \\
& \leq \frac{1}{2} \eta_i^{-2} \Phi_i^T(y) \Phi_i(y) + 2\eta_i^2 w_i^T M_i^* G_i G_i^T M_i^* w_i, \\
& \times w_i^T M_i^* K_i B_i \Psi_i(y) + \Psi_i^T(y) B_i^T K_i^T M_i^* w_i \\
& \leq \frac{1}{2} (1 - \alpha_i) \gamma_i^{-2} \Psi_i^T(y) \Psi_i(y) \\
& \quad + 2(1 - \alpha_i)^{-1} \gamma_i^2 w_i^T M_i^* K_i B_i B_i^T K_i^T M_i^* w_i.
\end{aligned}
\tag{6.51}
$$

Substituting (6.51) into (6.49) leads to

$$
\begin{aligned}
\dot{V}_i \leq\ & \zeta_i^T \Xi_i \zeta_i - x_i^T P_i^* B_i \left(R_i^{-1} - \frac{2\gamma_i^2}{\alpha_i} I_{m_i}\right) B_i^T P_i^* x_i \\
& - \frac{\epsilon_i}{h_i} w_i^T w_i - w_i^T M_i^* (G_i S_i^{-1} G_i^T - \frac{1}{\epsilon_i} \bar{E}_i \bar{E}_i^T - 2\eta_i^2 G_i G_i^T \\
& - 2(1 - \alpha_i)^{-1} \gamma_i^2 K_i B_i B_i^T K_i^T) M_i^* w_i - \left(1 + \frac{1}{h_i}\right) y_i^T y_i
\end{aligned}
$$

$$-\frac{\epsilon_i}{h_i} x_i^T x_i + \frac{1}{2}\gamma_i^{-2}\Psi_i^T(y)\Psi_i(y) + \frac{1}{2}\eta_i^{-2}\Phi_i^T(y)\Phi_i(y)$$

$$\leq -\delta_i \zeta_i^T \zeta_i. \tag{6.52}$$

From the conditions in Theorem 6.3.2, it can be derived that

$$\dot{V}_i \leq -h_i^{-1}|y_i|^2 - \epsilon_i h_i^{-1}|\zeta_i|^2 + \sum_{j=1, j\neq i}^{N} |y_j|^2. \tag{6.53}$$

This completes the proof.   ∎

Similarly, a sufficient result for the global asymptotic stability of the closed-loop large-scale system (6.34)–(6.36) and (6.43) is given in the following theorem, which can be proved using the same reasoning as in the proof of Theorem 6.1.3.

Theorem 6.3.3   *Under the conditions of Theorem 6.3.2, suppose the following inequality holds*

$$\sum_{j=1}^{N} \frac{h_j}{1+h_j} < 1 \tag{6.54}$$

Then, there exist constants $c_i > 0$, $i = 1, 2, \ldots, N$, such that

$$\frac{d}{dt}\sum_{i=1}^{N}\left(x_i^T c_i h_i P_i^* x_i + w_i^T c_i h_i M_i^* w_i\right) \leq -|y|^2 - \sum_{i=1}^{N} c_i \epsilon_i |\zeta_i|^2 \tag{6.55}$$

Remark 6.3.4   *Note that the interactions among the subsystems can also be studied using the recently developed cyclic-small-gain condition [10, 20]. In that case, the assumptions $|\Psi_i(y)| \leq \gamma_i|y|$ and $|\Phi_i(y)| \leq \eta_i|y|$ can be relaxed, and similar stability results based on the input-to-state stability (ISS) property [30–32] can be obtained.*

Remark 6.3.5   *With appropriate selection of the weighting matrices, RADP-based online implementation for solving the robust optimal controller can be achieved using the same two-phase-learning schemes introduced in Chapter 5.*

6.4   APPLICATION TO A TEN-MACHINE POWER SYSTEM

6.4.1   Mathematical Model of a Multimachine Power System

Consider the classical multimachine power system with governor controllers [15]

$$\dot{\delta}_i(t) = \omega_i(t), \tag{6.56}$$

$$\dot{\omega}_i(t) = -\frac{D_i}{2H_i}\omega_i(t) + \frac{\omega_0}{2H_i}[P_{mi}(t) - P_{ei}(t)], \tag{6.57}$$

$$\dot{P}_{mi}(t) = \frac{1}{T_i}[-P_{mi}(t) + u_{gi}(t)], \tag{6.58}$$

$$P_{ei}(t) = E'_{qi}\sum_{j=1}^{N} E'_{qj}[B_{ij}\sin\delta_{ij}(t) + G_{ij}\cos\delta_{ij}(t)] \tag{6.59}$$

where $\delta_i(t)$ is the angle of the ith generator, $\delta_{ij} = \delta_i - \delta_j$; $\omega_i(t)$ is the relative rotor speed; $P_{mi}(t)$ and $P_{ei}(t)$ are the mechanical power and the electrical power; E'_{qi} is the transient electromotive force (EMF) in quadrature axis, and is assumed to be constant under high-gain silicon controlled rectifier (SCR) controllers; D_i, H_i, and T_i are the damping constant, the inertia constant, and the governor time constant; B_{ij}, G_{ij} are constants for $1 \le i, j \le N$.

Similarly as in [7], system (6.56)–(6.58) can be put into the following form,

$$\Delta\dot{\delta}_i(t) = \Delta\omega_i(t), \tag{6.60}$$

$$\Delta\dot{\omega}_i(t) = -\frac{D_i}{2H_i}\Delta\omega_i(t) + \frac{\omega_0}{2H_i}\Delta P_{mi}(t), \tag{6.61}$$

$$\Delta\dot{P}_{mi}(t) = \frac{1}{T_i}[-\Delta P_{mi}(t) + u_i(t) - d_i(t)] \tag{6.62}$$

where

$$\Delta\delta_i(t) = \delta_i(t) - \delta_{i0},$$

$$\Delta\omega_i(t) = \omega_i(t) - \omega_{i0},$$

$$\Delta P_{mi}(t) = P_{mi}(t) - P_{ei}(t),$$

$$u_i(t) = u_{gi}(t) - P_{ei}(t),$$

$$d_i(t) = E'_{qi}\sum_{j=1,j\neq i}^{N} E'_{qj}[B_{ij}\cos\delta_{ij}(t) - G_{ij}\sin\delta_{ij}(t)][\Delta\omega_i(t) - \Delta\omega_j(t)].$$

Assume there exists a constant $\beta > 0$ such that $\max\limits_{1\le i,j\le N}[E'_{qi}E'_{qj}(|B_{ij}| + |G_{ij}|) < \beta$. Then,

$$|d_i(t)| \le (N-1)\beta\sum_{j=1}^{N}|\Delta\omega_i - \Delta\omega_j| \le (N-1)^2\beta\sum_{j=1}^{N}|\Delta\omega_j|$$

Therefore, the model (6.60)–(6.62) is in the form (6.1) and (6.2), if we define $x_i = [\Delta\delta_i(t)\ \Delta\omega_i(t)\ \Delta P_{ei}(t)]^T$ and $y_i = \Delta\omega_i(t)$.

TABLE 6.1 Parameters for the Generators

	G1	G2	G3	G4	G5	G6	G7	G8	G9	G10
H_i (p.u.)	∞	6.4	3	5.5	5.2	4.7	5.4	4.9	5.1	3.4
D_i (p.u.)	—	1	1.5	2	2.2	2.3	2.6	1.8	1.7	2.9
T_i (s)	—	6	6.3	4.9	6.6	5.8	5.9	5.5	5.4	5.5
E_{qi} (p.u.)	1	1.2	1.5	0.8	1.3	0.9	1.1	0.6	1.5	1
$\delta_{i0}(°)$	0	108.86	97.4	57.3	68.75	74.48	45.84	68.75	40.11	63.03

6.4.2 Numerical Simulation

A ten-machine power system is considered for numerical studies. In the simulation, Generator 1 is used as the reference machine. Governor controllers and ADP-based learning systems are installed on Generators 2–10.

Simulation parameters for the ten-machine power system are shown in Tables 6.1– 6.3. Also the steady-state frequency is set to be $\omega_0 = 314.15$rad/s. The initial

TABLE 6.2 Imaginary Parts of the Admittance Matrix

B_{ij}	$j = 1$	$j = 2$	$j = 3$	$j = 4$	$j = 5$	$j = 6$	$j = 7$	$j = 8$	$j = 9$	$j = 10$
$i = 1$	0.25	0.19	0.41	0.30	0.29	0.48	0.24	0.09	0.21	0.23
$i = 2$	0.19	0.39	0.25	0.53	0.28	0.29	0.38	0.33	0.24	0.38
$i = 3$	0.41	0.25	0.05	0.27	0.25	0.22	0.27	0.21	0.33	0.49
$i = 4$	0.30	0.53	0.27	0.57	0.33	0.33	0.13	0.29	0.49	0.13
$i = 5$	0.29	0.28	0.25	0.32	0.21	0.37	0.30	0.03	0.24	0.31
$i = 6$	0.48	0.29	0.22	0.33	0.37	0.46	0.28	0.41	0.35	0.14
$i = 7$	0.24	0.38	0.27	0.13	0.30	0.28	0.12	0.43	0.14	0.53
$i = 8$	0.09	0.33	0.21	0.29	0.03	0.41	0.43	0.33	0.16	0.44
$i = 9$	0.21	0.24	0.33	0.49	0.24	0.35	0.14	0.16	0.36	0.21
$i = 10$	0.23	0.38	0.49	0.13	0.31	0.14	0.53	0.44	0.21	0.37

TABLE 6.3 Real Parts of the Admittance Matrix

G_{ij}	$j = 1$	$j = 2$	$j = 3$	$j = 4$	$j = 5$	$j = 6$	$j = 7$	$j = 8$	$j = 9$	$j = 10$
$i = 1$	0.21	0.05	0.00	0.14	0.04	−0.01	0.00	0.28	−0.08	−0.00
$i = 2$	0.05	0.29	0.03	0.02	0.13	−0.17	−0.00	−0.01	−0.11	−0.04
$i = 3$	0.00	0.03	0.15	0.18	−0.27	−0.10	0.15	−0.05	0.23	0.28
$i = 4$	0.13	0.01	0.18	0.05	−0.11	−0.07	0.22	−0.11	0.24	−0.15
$i = 5$	0.04	0.13	−0.27	−0.11	0.08	−0.20	−0.07	−0.19	0.03	−0.15
$i = 6$	−0.01	−0.17	−0.10	−0.07	−0.20	0.02	−0.03	0.04	0.12	−0.07
$i = 7$	0.00	−0.00	0.15	0.22	−0.07	−0.03	0.04	0.06	0.06	−0.10
$i = 8$	0.28	−0.01	−0.05	−0.11	−0.19	0.04	0.06	0.27	0.13	0.11
$i = 9$	−0.08	−0.11	0.23	0.23	0.03	0.12	0.06	0.13	0.19	−0.17
$i = 10$	−0.00	−0.04	0.28	−0.15	−0.15	−0.07	−0.10	0.11	−0.17	0.21

feedback policies are

$$K_i^{(0)} = [10 \quad 50 \quad 0], 1 \le i \le 10 \tag{6.63}$$

All the parameters, except for the operating points, are assumed to be unknown to the learning system. The weighting matrices are set to be $Q_i = 1000I_3$, $R_i = 1$, for $i = 2, 3, \dots, 10$.

From $t = 0s$ to $t = 1s$, all the generators were operating at the steady state. At $t = 1s$, an impulse disturbance on the active power was added to the network. As a result, the power angles and frequencies started to oscillate. In order to stabilize the system and improve its performance, the learning algorithm is conducted from $t = 4s$ to $t = 5s$. Robust ADP-based control policies for the generators are applied from $t = 5s$ to the end of the simulation. Trajectories of the angles and frequencies of Generators 2–4 are shown in Figures 6.2 and 6.3.

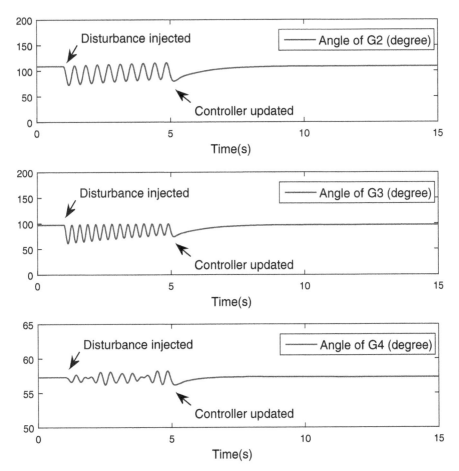

FIGURE 6.2 Rotor angle deviations of Generators 2–4. *Source*: Jiang, 2012. Reproduced with permission of IEEE.

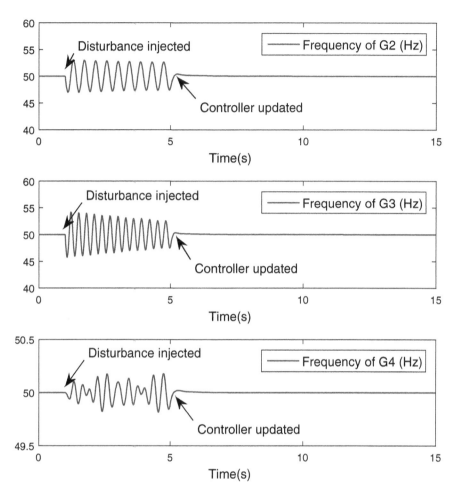

FIGURE 6.3 Rotor angle deviations of Generators 2–4. *Source*: Jiang, 2012. Reproduced with permission of IEEE.

6.5 NOTES

The development of intelligent online learning controller design methods gains remarkable popularity in the operation of large-scale complex systems, such as power systems. In recent years, considerable attention has been paid to the stabilization of large-scale complex systems [8, 22, 25, 27, 28], as well as the related consensus and synchronization problems [2, 18, 29, 35]. Examples of large-scale systems arise from ecosystems, transportation networks, and power systems, to name only a few, [7, 17, 21, 23, 34, 37]. Often, in real-world applications, precise mathematical models are hard to build and the model mismatches, caused by parametric and dynamic uncertainties, are thus unavoidable. This, together with the exchange of only local system information, makes the design problem extremely challenging in the context of complex networks.

This chapter extends the decentralized control of large-scale complex systems with uncertain system dynamics. RADP-based online learning method is developed for a class of large-scale systems, and a novel decentralized controller implementation algorithm is presented. The obtained controller globally asymptotically stabilizes the large-scale system, and at the same time, preserves suboptimality properties. In addition, the effectiveness of the proposed methodology is demonstrated via its application to the online learning control of a ten-machine power system with governor controllers. The extension to the unmatched case is addressed, and more details can be found in [1]. The approaches presented in this chapter are applicable to a large variety of complex real-world power networks [3–5, 24, 33, 36], wind conversion systems [26], and demand response management problems [6].

REFERENCES

[1] T. Bian, Y. Jiang, and Z. P. Jiang. Decentralized adaptive optimal control of large-scale systems with application to power systems. *IEEE Transactions on Industrial Electronics*, 62(4):2439–2447, 2015.

[2] M. Chen. Some simple synchronization criteria for complex dynamical networks. *IEEE Transactions on Circuits and Systems II: Express Briefs*, 53(11):1185–1189, 2006.

[3] P.-C. Chen and M. Kezunovic. Analysis of the impact of distributed generation placement on voltage profile in distribution systems. In: Proceedings of the 2013 IEEE Power and Energy Society General Meeting (PES), pp. 1–5, Vancouver, BC, July 2013.

[4] P.-C. Chen, V. Malbasa, and M. Kezunovic. Analysis of voltage stability issues with distributed generation penetration in distribution networks. In: Proceedings of the 2013 IEEE North American Power Symposium (NAPS), pp. 1–6, Manhattan, KS, 2013.

[5] P.-C. Chen, R. Salcedo, Q. Zhu, F. de León, D. Czarkowski, Z. P. Jiang, V. Spitsa, Z. Zabar, and R. E. Uosef. Analysis of voltage profile problems due to the penetration of distributed generation in low-voltage secondary distribution networks. *IEEE Transactions on Power Delivery*, 27(4):2020–2028, 2012.

[6] Z. Chen, L. Wu, and Y. Fu. Real-time price-based demand response management for residential appliances via stochastic optimization and robust optimization. *IEEE Transactions on Smart Grid*, 3(4):1822–1831, 2012.

[7] G. Guo, Y. Wang, and D. J. Hill. Nonlinear output stabilization control for multimachine power systems. *IEEE Transactions on Circuits and Systems I: Fundamental Theory and Applications*, 47(1):46–53, 2000.

[8] Z. P. Jiang. Decentralized control for large-scale nonlinear systems: A review of recent results. *Dynamics of Continuous, Discrete and Impulsive Systems*, 11:537–552, 2004.

[9] Z. P. Jiang, A. R. Teel, and L. Praly. Small-gain theorem for ISS systems and applications. *Mathematics of Control, Signals and Systems*, 7(2):95–120, 1994.

[10] Z. P. Jiang and Y. Wang. A generalization of the nonlinear small-gain theorem for large-scale complex systems. In: Proceedings of the 7th IEEE World Congress on Intelligent Control and Automation (WCICA), pp. 1188–1193, Chongqing, China, 2008.

[11] I. Karafyllis and Z. P. Jiang. A vector small-gain theorem for general non-linear control systems. *IMA Journal of Mathematical Control and Information*, 28(3):309–344, 2011.

[12] H. K. Khalil. *Nonlinear Systems*, 3rd ed. Prentice Hall, Upper Saddle River, NJ, 2002.

[13] D. Kleinman. On an iterative technique for Riccati equation computations. *IEEE Transactions on Automatic Control*, 13(1):114–115, 1968.

[14] M. Krstic, I. Kanellakopoulos, and P. V. Kokotovic. *Nonlinear and Adaptive Control Design*. John Wiley & Sons, New York, 1995.

[15] P. Kundur, N. J. Balu, and M. G. Lauby. *Power System Stability and Control*, Vol. 7. McGraw-Hill, New York, 1994.

[16] F. Lewis, D. Vrabie, and V. Syrmos. *Optimal Control*, 3rd ed. John Wiley & Sons, Inc., Hoboken, NJ, 2012.

[17] Z. Li and G. Chen. Global synchronization and asymptotic stability of complex dynamical networks. *IEEE Transactions on Circuits and Systems II: Express Briefs*, 53(1):28–33, 2006.

[18] Z. Li, Z. Duan, G. Chen, and L. Huang. Consensus of multiagent systems and synchronization of complex networks: A unified viewpoint. *IEEE Transactions on Circuits and Systems I: Regular Papers*, 57(1):213–224, 2010.

[19] T. Liu, D. J. Hill, and Z. P. Jiang. Lyapunov formulation of ISS cyclic-small-gain in continuous-time dynamical networks. *Automatica*, 47(9):2088–2093, 2011.

[20] T. Liu, Z. P. Jiang, and D. J. Hill. *Nonlinear Control of Dynamic Networks*. CRC Press, 2014.

[21] Y. Liu, T. Chen, C. Li, Y. Wang, and B. Chu. Energy-based disturbance L_2 attenuation excitation control of differential algebraic power systems. *IEEE Transactions on Circuits and Systems II: Express Briefs*, 55(10):1081–1085, 2008.

[22] A. N. Michel. On the status of stability of interconnected systems. *IEEE Transactions on Automatic Control*, 28(6):639–653, 1983.

[23] G. Revel, A. E. Leon, D. M. Alonso, and J. L. Moiola. Bifurcation analysis on a multimachine power system model. *IEEE Transactions on Robust Nonlinear Coordinated Control of Power Systems Circuits and Systems I: Regular Papers*, 57(4):937–949, 2010.

[24] R. Salcedo, X. Ran, F. de León, D. Czarkowski, and V. Spitsa. Long duration overvoltages due to current backfeeding in secondary networks. *IEEE Transactions on Power Delivery*, 28(4):2500–2508, October 2013.

[25] N. R. Sandell, P. Varaiya, M. Athans, and M. G. Safonov. Survey of decentralized control methods for large scale systems. *IEEE Transactions on Automatic Control*, 23(2):108–128, 1978.

[26] Y. She, X. She, and M. E. Baran. Universal tracking control of wind conversion system for purpose of maximum power acquisition under hierarchical control structure. *IEEE Transactions on Energy Conversion*, 26(3):766–775, 2011.

[27] D. D. Siljak. *Large-Scale Dynamic Systems: Stability and Structure*, Vol. 310. North-Holland, New York, 1978.

[28] D. D. Šiljak. *Decentralized Control of Complex Systems*. Academic Press, 1991.

[29] Q. Song and J. Cao. On pinning synchronization of directed and undirected complex dynamical networks. *IEEE Transactions on Circuits and Systems I: Regular Papers*, 57(3):672–680, 2010.

[30] E. D. Sontag. On the input-to-state stability property. *European Journal of Control*, 1(1):24–36, 1995.

[31] E. D. Sontag. Input to state stability: Basic concepts and results. In: *Nonlinear and Optimal Control Theory*, pp. 163–220. Springer, 2008.

[32] E. D. Sontag and Y. Wang. On characterizations of the input-to-state stability property. *Systems & Control Letters*, 24(5):351–359, 1995.

[33] V. Spitsa, X. Ran, R. Salcedo, J. F. Martinez, R. E. Uosef, F. de León, D. Czarkowski, and Z. Zabar. On the transient behavior of large-scale distribution networks during automatic feeder reconfiguration. *IEEE Transactions on Smart Grid*, 3(2):887–896, 2012.

[34] Y. Wang and D. J. Hill. Robust nonlinear coordinated control of power systems. *Automatica*, 32(4):611–618, 1996.

[35] X. Yang, J. Cao, and J. Lu. Stochastic synchronization of complex networks with non-identical nodes via hybrid adaptive and impulsive control. *IEEE Transactions on Circuits and Systems I: Regular Papers*, 59(2):371–384, 2012.

[36] L. Yu, D. Czarkowski, and F. de León. Optimal distributed voltage regulation for secondary networks with DGs. *IEEE Transactions on Smart Grid*, 3(2):959–967, 2012.

[37] J. Zhou and Y. Ohsawa. Improved swing equation and its properties in synchronous generators. *IEEE Transactions on Circuits and Systems I: Regular Papers*, 56(1):200–209, 2009.

CHAPTER 7

ROBUST ADAPTIVE DYNAMIC PROGRAMMING AS A THEORY OF SENSORIMOTOR CONTROL

Many tasks that humans perform in our everyday life involve different sources of uncertainties. However, it is interesting and surprising to notice how the central nervous system (CNS) can gracefully coordinate our movements to deal with these uncertainties. For example, one might be clumsy in moving an object with uncertain mass and unknown friction at the first time, but after several trials, the movements will gradually become smooth. Although extensive research by many authors has been made, the underlying computational mechanisms in sensorimotor control require further investigations.

This chapter studies sensorimotor control with static and dynamic uncertainties under the framework of RADP, as introduced in Chapters 5 and 6. In this chapter, the linear version of RADP is extended for stochastic systems by taking into account signal-dependent noise [17], and the proposed method is applied to study the sensorimotor control problem with both static and dynamic uncertainties (see Figure 7.1 for a related experimental setup). Results presented in this chapter suggest that the CNS may use RADP-like learning strategy to coordinate movements and to achieve successful adaptation in the presence of static and/or dynamic uncertainties. In the absence of dynamic uncertainties, the learning strategy reduces to an ADP-like mechanism.

Robust Adaptive Dynamic Programming, First Edition. Yu Jiang and Zhong-Ping Jiang.
© 2017 by The Institute of Electrical and Electronics Engineers, Inc. Published 2017 by John Wiley & Sons, Inc.

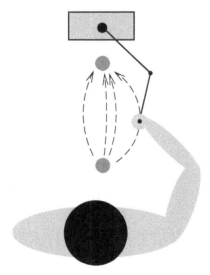

FIGURE 7.1   Top-down view of a seated human subject grasping the handle of a mechanical device that measures the hand position and generates disturbance forces.

7.1 ADP FOR CONTINUOUS-TIME STOCHASTIC SYSTEMS

7.1.1 Problem Formulation

To study sensorimotor control, we consider the following system governed by a stochastic differential equation

$$dx = Axdt + Budt + B\Sigma_{i=1}^{q} C_i u d\eta_i \qquad (7.1)$$

where $A \in \mathbb{R}^{n \times n}$ and $B \in \mathbb{R}^{n \times m}$ are constant matrices describing the system dynamics with the pair (A, B) stabilizable, $u \in \mathbb{R}^m$ is the control signal, η_i are independent scalar Brownian motions, and $C_i \in \mathbb{R}^{m \times m}$ are constant matrices, for $i = 1, 2, ..., q$.

The control objective is to determine a linear control policy

$$u = -Kx \qquad (7.2)$$

which minimizes the following cost

$$J(x_0; u) = \int_0^{\infty} (x^T Q x + u^T R u) dt \qquad (7.3)$$

for the nominal deterministic system of (7.1) (i.e., system (7.1) with $\eta_i = 0$, $\forall i = 1, 2, ..., q$), where $Q = Q^T \geq 0$, $R = R^T > 0$, with $(A, Q^{1/2})$ observable.

As mentioned in Chapter 2, when both A and B are accurately known, solution to this problem can be found by solving the following well-known algebraic Riccati equation (ARE)

$$A^T P + PA + Q - PBR^{-1}B^T P = 0 \tag{7.4}$$

which has a unique symmetric positive definite solution $P^* \in \mathbb{R}^{n \times n}$. In addition, the optimal feedback gain matrix K^* in (7.2) can thus be determined by

$$K^* = R^{-1}B^T P^*. \tag{7.5}$$

In the presence of the signal-dependent noise η_i, the closed-loop system is mean-square stable [30], if the following matrix is Hurwitz

$$(A - BK^*) \otimes I_n + I_n \otimes (A - BK^*) + \Sigma_{i=1}^q \left(BC_i K^* \otimes BC_i K^* \right). \tag{7.6}$$

When the constant matrices $C_i, i = 1, 2, \ldots, q$ are so small that (7.6) is Hurwitz, the control policy $u = -K^* x$ is called *robust optimal*, that is, it is optimal in the absence of the noise ζ_i, and is stabilizing in the presence of ζ_i.

7.1.2 Policy Iteration

The ARE in (7.4) can be solved using the same policy iteration algorithm as mentioned in Chapter 2. For the reader's convenience, the algorithm is rewritten as follows.

Algorithm 7.1.1   *Conventional policy iteration algorithm [29]*

(1) *Initialization:*
   Find an initial stabilizing feedback gain matrix K_0, such that $A - BK_0$ is Hurwitz.

(2) *Policy evaluation:*
   Solve P_k from

$$0 = (A - BK_k)^T P_k + P_k (A - BK_k) + Q + K_k^T R K_k \tag{7.7}$$

(3) *Policy improvement:*
   Improve the control policy by

$$K_{k+1} = R^{-1}B^T P_k \tag{7.8}$$

Then, go to Step (2) and solve for P_{k+1} with K_k replaced by K_{k+1}.

Again, it has been shown in [29] that the sequences $\{P_k\}$ and $\{K_k\}$ iteratively determined from policy iteration (7.7) and (7.8) have the following properties:

(1) $A - BK_k$ is Hurwitz,

(2) $P^* \le P_{k+1} \le P_k$, and

(3) $\lim_{k \to \infty} K_k = K^*$, $\lim_{k \to \infty} P_k = P^*$.

7.1.3 ADP for Linear Stochastic Systems with Signal-Dependent Noise

The policy iteration algorithm relies on the perfect knowledge of the system dynamics, because the system matrices A and B are involved in the equations (7.7) and (7.8). In Chapter 2, we have shown that in the deterministic case, when A and B are unknown, equivalent iterations can be achieved using online measurements. Here we extend the methodology introduced in Chapter 2 to deal with stochastic linear systems with signal-dependent noise, and to find online the optimal control policy without assuming the a priori knowledge of A and B.

To begin with, let us rewrite the original system (7.1) as

$$dx = (A - BK_k)xdt + B(dw + K_kxdt) \tag{7.9}$$

where

$$dw = udt + \Sigma_{i=1}^{q} C_i u d\eta_i \tag{7.10}$$

represents the combined signal received by the motor plant from the input channel.

Now, let us define $A_k = A - BK_k$, $Q_k = Q + K_k^T R K_k$, and $M_k = B^T P_k B$. Then, by Itô's lemma [22], along the solutions of (7.9), it follows that

$$
\begin{aligned}
&d\left(x^T P_k x\right) \\
&= dx^T P_k x + x^T P_k dx + dx^T P_k dx \\
&= x^T \left(A_k^T P_k + P_k A_k\right) xdt + 2(K_k xdt + dw)^T B^T P_k x \\
&\quad + dw^T B^T P_k B dw \\
&= -x^T Q_k xdt + 2(K_k xdt + dw)^T B^T P_k x + dw^T M_k dw
\end{aligned} \tag{7.11}
$$

Notice that $dw^T M_k dw \ne 0$ because

$$
\begin{aligned}
&dw^T M_k dw \\
&= \left(udt + \Sigma_{i=1}^{q} C_i u d\eta_i\right)^T M_k \left(udt + \Sigma_{i=1}^{q} C_i u d\eta_i\right) \\
&= u^T M_k u(dt)^2 + \Sigma_{i=1}^{q} u^T C_i^T M_k C_i u(d\eta_i)^2 \\
&\quad + \Sigma_{1 \le i \ne j \le q} u^T C_i^T M_k C_j u d\eta_i d\eta_j \\
&= \Sigma_{i=1}^{q} u^T C_i^T M_k C_i u dt
\end{aligned}
$$

Next, integrating both sides of (7.11) from t to $t + \delta t$, we obtain

$$
x(t + \delta t)^T P_k x(t + \delta t) - x(t)^T P_k x(t)
$$
$$
= - \int_t^{t+\delta t} \left(x^T Q x + u_k^T R u_k \right) dt + \int_t^{t+\delta t} dw^T M_k dw \qquad (7.12)
$$
$$
+ 2 \int_t^{t+\delta t} (K_k x + dw)^T R K_{k+1} x
$$

where $u_k = -K_k x$.

We now show that given a matrix K_k such that $A - BK_k$ is Hurwitz, a pair of matrices (P_k, K_{k+1}), with $P_k = P_k^T > 0$, satisfying (7.7) and (7.8) can be uniquely determined without knowing A or B. To this end, recall that by Kronecker product representation [20], we have

$$
(K_k x dt + dw)^T R K_{k+1} x = [x \otimes R(K_k x dt + dw)]^T \text{vec}(K_{k+1}) \qquad (7.13)
$$

Hence, for a sufficiently large positive integer $l_k > 0$, we define matrices $\Theta_k \in \mathbb{R}^{l_k \times (n^2 + m^2 + mn)}$ and $\Xi_k \in \mathbb{R}^{l_k}$, such that

$$
\Theta_k = \begin{bmatrix}
x^T \otimes x^T \big|_{t_{0,k}}^{t_{1,k}} & -\int_{t_{0,k}}^{t_{1,k}} dw^T \otimes dw^T & -2 \int_{t_{0,k}}^{t_{1,k}} x^T \otimes (K_k x dt + dw)^T R \\
x^T \otimes x^T \big|_{t_{1,k}}^{t_{2,k}} & -\int_{t_{1,k}}^{t_{2,k}} dw^T \otimes dw^T & -2 \int_{t_{1,k}}^{t_{2,k}} x^T \otimes (K_k x dt + dw)^T R \\
\vdots & \vdots & \vdots \\
x^T \otimes x^T \big|_{t_{l_k-1,k}}^{t_{l_k,k}} & -\int_{t_{l_k-1,k}}^{t_{l_k,k}} dw^T \otimes dw^T & -2 \int_{t_{l_k-1,k}}^{t_{l_k,k}} x^T \otimes (K_k x dt + dw)^T R
\end{bmatrix},
$$

$$
\Xi_k = - \begin{bmatrix}
\int_{t_{0,k}}^{t_{1,k}} \left(x^T Q x + u_k^T R u_k \right) dt \\
\int_{t_{1,k}}^{t_{2,k}} \left(x^T Q x + u_k^T R u_k \right) dt \\
\cdots \\
\int_{t_{l_k-1,k}}^{t_{l_k,k}} \left(x^T Q x + u_k^T R u_k \right) dt
\end{bmatrix},
$$

where $0 \le t_{l_{k-1},k-1} \le t_{0,k} < t_{1,k} < \cdots < t_{l_k,k} < t_{0,k+1}$.

Therefore, (7.12) implies the following compact form of linear equations

$$
\Theta_k \begin{bmatrix} \text{vec}(P_k) \\ \text{vec}(M_k) \\ \text{vec}(K_{k+1}) \end{bmatrix} = \Xi_k \qquad (7.14)
$$

Further, to guarantee the existence and uniqueness of solution to (7.14), similar as in Chapter 2, we assume the following rank condition:

$$
\text{rank}(\Theta_k) = \frac{n(n+1) + m(m+1)}{2} + mn, \forall k \in \mathbb{Z}_+ \qquad (7.15)
$$

Alternatively, equation (7.14) can be replaced with recursive least-squares (RLS) methods [37].

7.1.4 The ADP Algorithm

Now, we are ready to summarize the ADP algorithm for practical online implementation.

Algorithm 7.1.2 *On-policy ADP algorithm with signal-dependent noise*

(1) *Initialization:*
Find an initial stabilizing control policy $u_0 = -K_0 x$, and set $k = 0$.
(2) *Online data collection:*
Apply $u_k = -K_k x$ as the control input on the time interval $[t_{0,k}, t_{l_k,k}]$. Compute Θ_k and Ξ_k.
(3) *Policy evaluation and improvement:*
Solve for $P_k = P_k^T > 0$, $M_k = M_k^T > 0$, and K_{k+1} from (7.14). Then, let $k \leftarrow k + 1$, and go to Step (2).

Compared with most of the existing models for motor adaptation, the proposed ADP algorithm can be used to study both the online learning during one single trial and the learning among different trials. In the latter case, the interval $[t_{0,k}, t_{l_k,k}]$ should be taken from the time duration of a single trial.

7.1.5 Convergence Analysis

The convergence property of the proposed algorithm can be summarized in the following theorem.

Theorem 7.1.3 *Suppose the rank condition (7.15) is satisfied and $A - BK_0$ is Hurwitz. Then, the sequences $\{K_k\}$, $\{P_k\}$, and $\{M_k\}$ obtained from (7.14) satisfy*
$$\lim_{k \to \infty} P_k = P^*, \quad \lim_{k \to \infty} K_k = K^*, \quad \text{and} \quad \lim_{k \to \infty} M_k = B^T P^* B.$$

Proof: Given a stabilizing feedback gain matrix K_k, if $P_k = P_k^T$ is the solution of (7.7), K_{k+1} and M_k are uniquely determined by $K_{k+1} = R^{-1} B^T P_k$ and $M_k = B^T P_k B$, respectively. By (7.12)–(7.14), we know that P_k, K_{k+1}, and M_k satisfy (7.14).
On the other hand, let $P = P^T \in \mathbb{R}^{n \times n}$, $M = M^T \in \mathbb{R}^{m \times m}$, and $K \in \mathbb{R}^{m \times n}$, such that

$$\Theta_k \begin{bmatrix} \text{vec}(P) \\ \text{vec}(M) \\ \text{vec}(K) \end{bmatrix} = \Xi_k. \tag{7.16}$$

Then, under the rank condition (7.15) and by the definition of Θ_k, we immediately know that the $\frac{n(n+1)+m(m+1)}{2} + mn$ independent entries in $\begin{bmatrix} \text{vec}(P) \\ \text{vec}(M) \\ \text{vec}(K) \end{bmatrix}$ are uniquely determined, indicating that $P = P^T$, $M = M^T$, and K are uniquely determined by (7.16). Hence, $P_k = P$, $M_k = M$, and $K_{k+1} = K$. Therefore, the pair (P_k, K_{k+1}) obtained from (7.14) is the same as the one obtained from (7.7) and (7.8). The convergence is thus proved. ∎

7.2 RADP FOR CONTINUOUS-TIME STOCHASTIC SYSTEMS

7.2.1 Problem Formulation

In this section, we generalize the commonly used linear models for sensorimotor control [12, 17, 23, 48] by taking into account the dynamic uncertainty, or unmodeled dynamics. To be more specific, consider the stochastic differential equations

$$dw = Fwdt + Gxdt \tag{7.17}$$

$$dx = Axdt + B_1 \left[zdt + \Delta_1(w,x)\,dt + \sum_{i=1}^{q_1} E_{1i}zd\eta_{1i} \right] \tag{7.18}$$

$$dz = B_2 \left[udt + \Delta_2(w,x,z)\,dt + \sum_{i=1}^{q_2} E_{2i}ud\eta_{2i} \right] \tag{7.19}$$

$$\Delta_1 = D_{11}w + D_{12}x \tag{7.20}$$

$$\Delta_2 = D_{21}w + D_{22}x + D_{23}z \tag{7.21}$$

where $[x^T, z^T]^T \in \mathbb{R}^{n+m}$ is the measurable state which represents the state of the sensorimotor system, $w \in \mathbb{R}^{n_w}$ is the unmeasurable state of the dynamic uncertainty, representing the unknown dynamics in the interactive environment, Δ_1 and Δ_2 are the outputs of the dynamic uncertainty, A, B_1, B_2, F, G, E_{1i} with $i = 1, 2,..., q_1$, and E_{2i} with $i = 1, 2,..., q_2$ are unknown constant matrices with suitable dimensions and $B_2 \in \mathbb{R}^{m \times m}$ is assumed to be invertible, η_{1i} with $i = 1, 2,..., q_1$ and η_{2i} with $i = 1, 2, ..., q_2$ are independent scalar Brownian motions, $u \in \mathbb{R}^m$ denotes the input of the motor control command.

The control design objective is to find a robust optimal feedback control policy which robustly stabilizes the overall system (7.17)–(7.21), and is optimal for the nominal system, that is, the system comprised of (7.18) and (7.19) with $\Delta_1 \equiv 0$, $\Delta_2 \equiv 0$, $E_{1i} = 0$, and $E_{2i} = 0$.

For this purpose, let us introduce an assumption on the dynamic uncertainty, which is modeled by the w-subsystem.

Assumption 7.2.1 *There exist $S = S^T > 0$ and a constant $\gamma_w > 0$, such that*

$$SF + F^T S + I + \gamma_w^{-1} SGG^T S < 0. \tag{7.22}$$

Remark 7.2.2   *Assumption 7.2.1 implies that the dynamic uncertainty, described by the w-subsystem, is finite-gain L_2 stable with a linear gain smaller than $\sqrt{\gamma_w}$, when x is considered as the input and w is considered as the output [32].*

7.2.2 Optimal Control Design for the (w, z)-Subsystem

Consider the subsystem comprised of (7.17), (7.18), and (7.20) with z regarded as the input. For convenience, the system is rewritten as follows

$$dw = Fwdt + Gxdt \tag{7.23}$$

$$dx = Axdt + B_1 \left[zdt + \Delta_1 (w, x) dt + \sum_{i=1}^{q_1} E_{1i} z d\eta_{1i} \right] \tag{7.24}$$

$$\Delta_1 = D_{11}w + D_{12}x \tag{7.25}$$

and the related nominal system is defined as

$$dx = Axdt + B_1 zdt \tag{7.26}$$

The cost associated with (7.26), with respect to z as the input, is selected to be

$$J_1(x_0; z) = \int_0^\infty \left(x^T Q_1 x + z^T R_1 z \right) dt \tag{7.27}$$

where $Q_1 = Q_1^T \geq 0$, $R_1 = R_1^T > 0$, and the pair $(A, Q_1^{1/2})$ is observable.
By linear optimal control theory [35], the optimal control policy takes the following form

$$z = -R_1^{-1} B_1^T P_1 x \tag{7.28}$$

where $P_1 = P_1^T > 0$ is the solution of the following ARE

$$A^T P_1 + P_1 A + Q_1 - P_1 B_1 R_1^{-1} B_1^T P_1 = 0. \tag{7.29}$$

To perform stability analysis, the following concept on mean-square stability [51] will be used in the remainder of this chapter.

Definition 7.2.3   *Consider the system*

$$dx = Axdt + \sum_{i=1}^{q} B_i x d\eta_i \tag{7.30}$$

where η_i with $i = 1, 2, \ldots, q$ are standard scalar Brownian motions. Then, the system is said to be stable in the mean-square sense if

$$\lim_{t \to \infty} E[x(t)x(t)^T] = 0. \tag{7.31}$$

Now, the following theorem gives stability criteria of the closed-loop system comprised of (7.23)–(7.25), and (7.28).

Theorem 7.2.4   *The closed-loop system comprised of (7.23), (7.24), (7.25), and (7.28) is mean-square stable if :*

1. *the weighting matrices Q_1 and R_1 are selected such that*

$$Q_1 > (\kappa_{12} + \kappa_{11}\gamma_w)I_n \quad \text{and} \quad R_1 < I_m \tag{7.32}$$

   where $\kappa_{11} = 2|D_{11}|^2$ and $\kappa_{12} = 2|D_{12}|^2$.
2. *the constant matrices E_{1i} with $i = 1, 2, \ldots, q_1$ satisfy*

$$\sum_{i=1}^{q_1} E_{1i}^T B_1^T P_1 B_1 E_{1i} \leq R_1(I_m - R_1) \tag{7.33}$$

Proof: First, we define $\mathcal{L}(\cdot)$ as the infinitesimal generator [34]. Then, along the trajectories of the x-subsystem, (7.24), we have

$$\mathcal{L}\left(x^T P_1 x\right) = -x^T \left(Q_1 + P_1 B_1 R^{-1} B_1^T P_1\right) x + 2x^T P_1 B_1 \Delta_1$$
$$+ x^T P_1 B_1 R_1^{-1} \sum_{i=1}^{q_1} E_{1i}^T B_1^T P_1 B_1 E_{1i} R_1^{-1} B_1^T P_1 x$$
$$= -x^T Q_1 x - \left|\Delta_1 - B_1^T P_1^T x\right|^2 dt + |\Delta_1|^2$$
$$- x^T P_1 B_1 R_1^{-1} \left[R_1 - R_1^2 - \sum_{i=1}^{q_1} E_{1i}^T B_1^T P_1 B_1 E_{1i}\right]$$
$$\times R_1^{-1} B_1^T P_1 x$$
$$\leq -x^T Q_1 x + |\Delta_1|^2$$

On the other hand, under Assumption 7.2.1, along the solutions of the w-subsystem (7.23), we have

$$\mathcal{L}(w^T S w) = w^T (SF + F^T S)w + w^T SGx + x^T G^T S w$$
$$< -|w|^2 - \gamma_w^{-1} w^T SGG^T S w + w^T SGx + x^T G^T S w$$
$$< -|w|^2 + \gamma_w |x|^2 \qquad (\forall w \neq 0)$$

By definition, we know that

$$
\begin{aligned}
|\Delta_1|^2 &= |D_{11}w + D_{12}x|^2 \\
&\leq 2|D_{11}|^2|w|^2 + 2|D_{12}|^2|x|^2 \\
&\leq \kappa_{11}|w|^2 + \kappa_{12}|x|^2
\end{aligned}
$$

Therefore, for all $(x, w) \neq 0$, the following holds

$$
\begin{aligned}
&\mathcal{L}\left(x^T P_1 x + \kappa_{11} w^T S w\right) \\
&< -\gamma_x |x|^2 + |\Delta_1|^2 - \kappa_{11}|w|^2 + \kappa_{11}\gamma_w|x|^2 \\
&< -\gamma_x |x|^2 + \kappa_{11}|w|^2 + 2x^T D_{12}^T D_{12}x - \kappa_{11}|w|^2 + \kappa_{11}\gamma_w|x|^2 \\
&< -x^T(Q_1 - \kappa_{12}I_n - \kappa_{11}\gamma_w I_n)x \\
&< 0
\end{aligned}
$$

Notice that $V(w, x) = x^T P_1 x + \kappa_{11} w^T S w$ can be regarded as a stochastic Lyapunov function [34], and the proof is thus complete. ∎

7.2.3 Optimal Control Design for the Full System

Now, we proceed ahead to study the full system. Define the transformation

$$
\xi = z + K_1 x \tag{7.34}
$$

where $K_1 = R_1^{-1}B_1^T P_1$. Then, we have

$$
\begin{aligned}
d\xi &= dz + K_1 dx \\
&= B_2\left[udt + \Delta_2 dt + \sum_{i=1}^{q_2} E_{2i}ud\eta_{2i}\right] \\
&\quad + K_1\left(A_c xdt + B_1\Delta_1 dt + B_1\xi dt + B_1\sum_{i=1}^{q_1} E_{1i}zd\eta_{1i}\right) \\
&= K_1 A_c xdt + B_2 udt + B_2\bar{\Delta}_2 dt + K_1 B_1\xi dt \\
&\quad + \sum_{i=1}^{q_2} B_2 E_{2i}ud\eta_{2i} + K_1\sum_{i=1}^{q_1} B_1 E_{1i}(\xi - K_1 x)d\eta_{1i}
\end{aligned}
$$

where $\bar{\Delta}_2 = B_2^{-1}K_1 B_1\Delta_1 + \Delta_2$ and $A_c = A - B_1 K_1$.

Consequently, the system (7.17)–(7.19) is converted to

$$
dw = Fwdt + Gxdt \tag{7.35}
$$

$$
dx = A_c xdt + B_1\left[\xi dt + \Delta_1 dt + \sum_{i=1}^{q_1} E_{1i}(\xi - K_1 x)d\eta_{1i}\right] \tag{7.36}
$$

$$
\begin{aligned}
d\xi &= K_1 A_c xdt + B_2 udt + B_2\bar{\Delta}_2 dt + K_1 B_1\xi dt \\
&\quad + \sum_{i=1}^{q_2} B_2 E_{2i}ud\eta_{2i} + K_1\sum_{i=1}^{q_1} B_1 E_{1i}(\xi - K_1 x)d\eta_{1i}
\end{aligned} \tag{7.37}
$$

Now, let us consider the control policy

$$u = -R_2^{-1} B_2^T P_2 \xi = -R_2^{-1} B_2^T P_2 \left(z + R_1^{-1} B_1 P_1 x \right) \tag{7.38}$$

where $P_2 = P_2^T > 0$ is the solution of the following equation

$$Q_2 - P_2 B_2 R_2^{-1} B_2^T P_2 = 0 \tag{7.39}$$

with Q_2 and R_2 two positive definite and symmetric matrices.

Remark 7.2.5 *Notice that (7.39) is an ARE associated with the following optimal control problem*

$$\min_u \quad J_2 = \int_0^\infty \left(z^T Q_2 z + u^T R_2 u \right) dt, \tag{7.40}$$

$$\text{s.t.} \quad \dot{z} = B_2 u, \tag{7.41}$$

where (7.41) is the nominal system of (7.19).

The stability criteria are given in the following theorem.

Theorem 7.2.6 *The closed-loop system comprised of (7.35)–(7.37) and (7.38) is mean-square stable if:*

1. *the weighting matrices Q_1 and R_1 as defined in (7.29), and Q_2 and R_2 as defined in (7.39) are selected such that $R_1 < I_m$, $R_2 < I_m$ and*

$$\begin{bmatrix} (\kappa_{12} + \kappa_{22} + \gamma_w \kappa_{11} + \gamma_w \kappa_{21})I_n & -A_c^T K_1^T P_2 - P_1 B_1 \\ -P_2 K_1 A_c - B_1^T P_1 & (\kappa_{23} + \kappa_3)I_m \end{bmatrix}$$

$$< \begin{bmatrix} Q_1 & 0 \\ 0 & Q_2 \end{bmatrix} \tag{7.42}$$

2. *the constant matrices E_{1i} and E_{2i} satisfy the following inequalities*

$$\sum_{i=1}^{q_1} E_{1i}^T B_1^T \left(P_1 + K_1^T P_2 K_1 \right) B_1 E_{1i} \le \frac{1}{2} R_1 (I_m - R_1) \tag{7.43}$$

$$\sum_{i=1}^{q_2} E_{2i}^T B_2^T P_2 B_2 E_{2i} \le R_2 (I_m - R_2) \tag{7.44}$$

where

$$\kappa_{21} = 3 \left| B_2^{-1} K_1 B_1 D_{11} + D_{21} \right|^2,$$

$$\kappa_{22} = 3 \left| B_2^{-1} K_1 B_1 D_{12} + D_{22} - D_{23} K_1^* \right|^2,$$

$$\kappa_{23} = 3 |D_{23}|^2,$$

$$\kappa_3 = \left| 2 \sum_{i=1}^{q_1} E_{1i}^T B_1^T K_1^T P_2 B_1 K_1 E_{1i} - P_2 K_1 B_1 - B_1^T K_1^T P_2 \right|.$$

Proof: Along the trajectories of the *x*-subsystem (7.18), we have

$$
\mathcal{L}\left(x^T P_1 x\right) \le -x^T Q_1 x - x^T P_1 B_1 \left(R_1^{-1} - I_m\right) B_1^T P_1 x
$$
$$
- x^T P_1 B_1 B_1^T P_1 x + 2x^T P_1 B_1 (\Delta_1 + \xi)
$$
$$
+ (\xi - K_1 x)^T \sum_{i=1}^{q_1} E_{1i}^T B_1^T P_1 B_1 E_{1i}(\xi - K_1 x)
$$
$$
\le -x^T Q_1 x + |\Delta_1|^2 + 2x^T P_1 B_1 \xi
$$
$$
- x^T P_1 B_1 \left(R^{-1} - I_m\right) B_1^T P_1 x
$$
$$
+ 2x^T K_1^T \sum_{i=1}^{q_1} E_{1i}^T B_1^T P_1 B_1 E_{1i} K_1 x
$$
$$
+ 2\xi^T \sum_{i=1}^{q_1} E_{1i}^T B_1^T P B_1 E_{1i} \xi
$$
$$
\le -x^T Q_1 x + |\Delta_1|^2 + 2x^T P_1 B_1 \xi
$$
$$
- x^T K_1^T \left[R_1 - R_1^2 - 2 \sum_{i=1}^{q_1} E_{1i}^T B_1^T P_1 B_1 E_{1i} \right] K_1 x
$$
$$
+ 2\xi^T \sum_{i=1}^{q_1} E_{1i}^T B_1^T P_1 B_1 E_{1i} \xi
$$

Then, along the trajectories of the *ξ*-subsystem,

$$
\mathcal{L}\left(\xi^T P_2 \xi\right) = 2\xi^T P_2 \left(K_1 A_c x + B_2 u + B_2 \bar{\Delta}_2 + K_1 B_1 \xi\right)
$$
$$
+ \sum_{i=1}^{q_1} (\xi - K_1 x)^T E_{1i}^T B_1^T K_1^T P_2 K_1 B_1 E_{1i}(\xi - K_1 x)
$$
$$
+ \sum_{i=1}^{q_2} \xi^T K_2^T E_{2i}^T B_2^T P_2 B_2 E_{2i} K_2 \xi
$$
$$
= -\xi^T \left(Q_2 - 2 \sum_{i=1}^{q_1} E_{1i}^T B_1^T K_1^T P_2 K_1 B_1 E_{1i} \right) \xi
$$
$$
+ \xi^T \left(P_2 K_1 B_1 + B_1^T K_1^T P_2 \right) \xi
$$
$$
+ 2\xi^T P_2 K_1 A_c x + |\bar{\Delta}_2|^2
$$
$$
+ 2 \sum_{i=1}^{q_1} x^T K_1^T E_{1i}^T B_1^T K_1^T P_2 K_1 B_1 E_{1i} K_1 x
$$
$$
- \sum_{i=1}^{q_2} \xi^T K_2^T \left(R_2 - R_2^2 - E_{2i}^T B_2^T P_2 B_2 E_{2i} \right) K_2 \xi
$$

Also, by definition

$$
\begin{aligned}
|\bar{\Delta}_2|^2 &= | \left(B_2^{-1} K_1 B_1 D_{11} + D_{21} \right) w \\
&\quad + \left(B_2^{-1} K_1 B_1 D_{12} + D_{22} \right) x \\
&\quad + D_{23} (\xi - K_1 x)|^2 \\
&= | \left(B_2^{-1} K_1 B_1 D_{11} + D_{21} \right) w \\
&\quad + \left(B_2^{-1} K_1 B_1 D_{12} + D_{22} - D_{23} K_1 \right) x \\
&\quad + D_{23} \xi|^2 \\
&\leq \kappa_{21} |w|^2 + \kappa_{22} |x|^2 + \kappa_{23} |\xi|^2
\end{aligned}
$$

Finally, along solutions of the closed-loop system (7.17)–(7.21) and (7.38), the following holds for all $[w^T, x^T, z^T] \neq 0$

$$
\begin{aligned}
& \mathcal{L} \left[(\kappa_{11} + \kappa_{21}) w^T S w + x^T P_1 x + \xi^T P_2 \xi \right]) \\
& \leq - \begin{bmatrix} x \\ \xi \end{bmatrix}^T \begin{bmatrix} Q_1 - (\kappa_{12} + \kappa_{22} + \kappa_{11}\gamma + \kappa_{21}\gamma) I_n & A_c^T K_1^T P_2 + P_1 B_1 \\ P_2 K_1 A_c + B_1^T P_1 & Q_2 - (\kappa_{23} + \kappa_3) I_m \end{bmatrix} \\
& \quad \times \begin{bmatrix} x \\ \xi \end{bmatrix} - x^T K_1^T \left(R_1 - R_1^2 \right) K_1 x \\
& \quad + 2 x^T K_1^T \sum_{i=1}^{q_1} E_{1i}^T B_1^T \left(P_1 + K_1^T P_2 K_1 \right) B_1 E_{1i} K_1 x \\
& \quad - \xi^T K_2^T \left(R_2 - R_2^2 - \sum_{i=1}^{q_2} E_{2i}^T B_2^T P_2 B_2 E_{2i} \right) K_2 \xi \\
& < 0
\end{aligned}
$$

The proof is complete.   ∎

Remark 7.2.7   *Consider the system comprised of (7.35)–(7.37). In the absence of dynamic uncertainties (i.e., the w-subsystem is absent, $\Delta_1 \equiv 0$, $\Delta_2 \equiv 0$, $E_{1i} = 0$ for $i = 1, 2, \ldots, q_1$, and $E_{2i} = 0$ for $i = 1, 2, \ldots, q_2$), the control policy (7.38) is optimal in the sense that it minimizes the cost*

$$
J_2 = \int_0^\infty \left(\begin{bmatrix} x \\ \xi \end{bmatrix}^T \bar{Q}_2 \begin{bmatrix} x \\ \xi \end{bmatrix} + u^T R_2 u \right) dt \tag{7.45}
$$

where

$$
\bar{Q}_2 = \begin{bmatrix} Q_1 + K_1^T R_1 K_1 & A_c^T K_1^T P_2 + P_1 B_1 \\ P_2 K_1 A_c + B_1^T P_1 & Q_2 - P_2 K_1 B_1 - B_1^T K_1^T P_2 \end{bmatrix} > 0
$$

Notice that the design methodology is different from inverse optimal control [33]. Indeed, although the weighing matrix $\bar{Q}_2$ cannot be arbitrarily specified, it can be indirectly modified by tuning the matrices Q_1, Q_2, R_1, and R_2. In motor control systems, Q_1, Q_2, and R_1 are related with the weights assigned on the movement accuracy, while R_2 represents the weights assigned on the control effort.

7.2.4 Offline Policy Iteration Technique

In order to obtain the robust optimal control policy (7.38), we need to first solve (7.29) and (7.39) which are nonlinear in P_1 and P_2, respectively. This can be done using the following offline policy iteration algorithm.

Algorithm 7.2.8   *Offline policy iteration algorithm*

(1) *Initialization:*
   Find feedback gain matrices $K_{1,0}$ and $K_{2,0}$, such that $A - B_1 K_{1,0}$ and $-B_2 K_{2,0}$ are both Hurwitz. Let $k = 0$ and

$$u_0 = -K_{2,0} z - K_{2,0} K_{1,0} x \qquad (7.46)$$

(2) *Phase-one policy evaluation:*
   Solve $P_{1,k}$ from

$$0 = (A - B_1 K_{1,k})^T P_{1,k} + P_{1,k}(A - B_1 K_{1,k}) \\ + Q_1 + K_{1,k}^{} R_1 K_{1,k} \qquad (7.47)$$

(3) *Phase-two policy evaluation:*
   Solve $P_{2,k}$ from

$$0 = -B_2 K_{2,k}^T P_{2,k} - P_{2,k} B_2 K_{2,k} + Q_2 + K_{2,k}^T R_2 K_{2,k} \qquad (7.48)$$

(4) *Policy improvement:*
   Let $k \leftarrow k + 1$, and improve the control policy by

$$u_k = -K_{2,k} z - K_{2,k} K_{1,k} x \qquad (7.49)$$

where

$$K_{i,k} = R_i^{-1} B_i^T P_{i,k-1}, \quad \forall i = 1, 2 \qquad (7.50)$$

Then, go to Step (2).

Remark 7.2.9   *The proposed method requires an initial stabilizing control policy. To be more specific, we need to know feedback gain matrices $K_{1,0}$ and $K_{2,0}$, such that*

$A - B_1 K_{1,0}$ and $-B_2 K_{2,0}$ are both Hurwitz. Even if A, B_1, and B_2 are uncertain, it is still possible to find such $K_{1,0}$ and $K_{2,0}$ by linear robust control theory, when some upper and lower bounds of the elements in A, B_1, and B_2 are available. In practice, these bounds can be estimated by the CNS during the first several trials. Take the model described by (7.64)–(7.66) as an example. In the absence of disturbances (i.e., $f \equiv 0$), we have

$$A - B_1 K_{1,0} = \begin{bmatrix} 0 & I_2 \\ 0 & -\frac{b}{m} I_2 \end{bmatrix} - \begin{bmatrix} 0 \\ \frac{1}{m} I_2 \end{bmatrix} K_{1,0} \tag{7.51}$$

and

$$B_2 K_{2,0} = \frac{1}{\tau} K_{2,0} \tag{7.52}$$

Since we know that $b > 0$, $m > 0$, and $\tau > 0$, we can choose, for example, $K_{1,0} = [I_2, 0]$ and $K_{2,0} = I_2$. Then, Algorithm 7.2.8 can proceed with the resultant initial stabilizing control policy.

Convergence of this offline policy iteration method can be concluded in the following theorem, of which the proof can be trivially extended from the proof of Theorem 2.1.1.

Theorem 7.2.10 *The sequences $\{P_{i,k}\}$, $\{K_{i,k}\}$ with $i = 1, 2$ and $k = 0, 1, \ldots$ iteratively determined from Algorithm 7.2.8 have the following properties, for $k = 0, 1, \ldots$*

(1) $A - B_1 K_{1,k}$ and $-B_2 K_{2,k}$ are both Hurwitz,
(2) $0 < P_i \leq P_{i,k+1} \leq P_{i,k}$, and
(3) $\lim_{k \to \infty} K_{i,k} = K_i$, $\lim_{k \to \infty} P_{i,k} = P_i$, $\forall i = 1, 2$.

7.2.5 Online Implementation

With the ADP methodology, the iteration steps in Algorithm 7.2.8 can be equivalently implemented using online sensory data without the need to identify the system dynamics. To begin with, let us rewrite the x-subsystem (7.18) as

$$dx = (A - B_1 K_{1,k})x dt + B_1(dw_1 + K_{1,k} x dt) \tag{7.53}$$

where

$$dw_1 = z dt + \Delta_1 dt + \Sigma_{i=1}^{q} E_{1i} z d\eta_{1i} \tag{7.54}$$

represents the combined signal received by the motor plant from the input channel.

By Itô's lemma [22], along the solutions of (7.53), it follows that

$$
\begin{aligned}
d\left(x^T P_{1,k} x\right) \\
&= dx^T P_{1,k} x + x^T P_{1,k} dx + dx^T P_{1,k} dx \\
&= x^T \left(A_k^T P_{1,k} + P_{1,k} A_k\right) x dt \\
&\quad + 2(K_{1,k} x dt + dw_1)^T B_1^T P_{1,k} x \\
&\quad + dw_1^T B_1^T P_{1,k} B_1 dw_1 \\
&= -x^T Q_{1,k} x dt \\
&\quad + 2(K_{1,k} x dt + dw_1)^T B_1^T P_{1,k} x \\
&\quad + dw_1^T M_{1,k} dw_1
\end{aligned}
\tag{7.55}
$$

where

$$
A_k = A - B_1 K_{1,k}, \tag{7.56}
$$

$$
Q_{1,k} = Q_1 + K_{1,k}^T R_1 K_{1,k}, \tag{7.57}
$$

$$
M_{1,k} = B_1^T P_{1,k} B_1. \tag{7.58}
$$

Next, integrating both sides of (7.55) from t to $t + \delta t$, we obtain

$$
\begin{aligned}
&x(t + \delta t)^T P_{1,k} x(t + \delta t) - x(t)^T P_{1,k} x(t) \\
&= - \int_t^{t+\delta t} x^T Q_{1,k} x dt + \int_t^{t+\delta t} dw_1^T M_{1,k} dw_1 \\
&\quad + 2 \int_t^{t+\delta t} (K_{1,k} x dt + dw_1)^T R_1 K_{1,k+1} x
\end{aligned}
\tag{7.59}
$$

In the previous section, we have shown that given a matrix $K_{1,k}$ such that $A - B_1 K_{1,k}$ is Hurwitz, a pair of matrices $(P_{1,k}, K_{1,k+1})$, with $P_{1,k} = P_{1,k}^T > 0$, satisfying (7.47) and (7.50) can be uniquely determined without knowing A or B_1, or both.

Similarly, for the z-subsystem, we have

$$
\begin{aligned}
&z(t + \delta t)^T P_{2,k} z(t + \delta t) - z(t)^T P_{2,k} z(t) \\
&= - \int_t^{t+\delta t} z^T Q_{2,k} z dt + \int_t^{t+\delta t} dw_2^T M_{2,k} dw_2 \\
&\quad + 2 \int_t^{t+\delta t} (K_{2,k} z dt + dw_2)^T R_2 K_{2,k+1} z
\end{aligned}
\tag{7.60}
$$

where

$$
Q_{2,k} = Q_2 + K_{2,k}^T R_2 K_{2,k}, \tag{7.61}
$$

$$
M_{2,k} = B_2^T P_{2,k} B_2, \tag{7.62}
$$

$$
dw_2 = u dt + \Delta_2 dt + \sum_{i=1}^{q_2} E_{2i} u d\eta_{2i}. \tag{7.63}
$$

The two online learning equations (7.59) and (7.60) can both be implemented in the same way as in Section 7.1.3, and rank conditions similar to (7.15) can be imposed to assure the uniqueness of the solution obtained during each iteration.

The sensorimotor learning algorithm based on the proposed RADP theory is thus summarized as follows.

Algorithm 7.2.11   *On-policy RADP algorithm for stochastic systems*

(1) *Initialization:*
    Apply an initial control policy in the form of (7.46), and let $k = 0$.
(2) *Online data collection:*
    Collect online sensory data and solve (7.59) and (7.60).
(3) *Policy evaluation and policy improvement:*
    Let $k \leftarrow k + 1$. Then, apply the new control policy (7.49) to the motor system and go to Step (2).

Remark 7.2.12   *It is worth noticing that the past measurements on dw_1 and dw_2 are assumed available for online learning purpose. In the following sections, we will see that they correspond to the combined control signals received by the muscles. These signals can be measured by the muscle spindle and the Golgi tendon organs, and can be transmitted to the brain via the peripheral nervous systems (PNS).*

7.3 NUMERICAL RESULTS: ADP-BASED SENSORIMOTOR CONTROL

7.3.1 Open-Loop Model of the Motor System

We adopted the proposed ADP algorithm to model arm movements in force fields, and to reproduce similar results observed from experiments [6, 13], in which each human subject performed a series of forward arm reaching movements in the horizontal plane. For simulation purpose, we used the mathematical model describing two-joint arm movements [36], as shown below

$$dp = vdt \tag{7.64}$$

$$mdv = (a - bv + f)dt \tag{7.65}$$

$$\tau da = (u - a)dt + d\xi \tag{7.66}$$

where $p = [p_x, p_y]^T$, $v = [v_x, v_y]^T$, $a = [a_x, a_y]^T$, $u = [u_x, u_y]^T$, $f = [f_x, f_y]$ are two-dimensional hand position, velocity, acceleration state, control signal, and external force generated from the field, respectively, m denotes the mass of the hand, b is the viscosity constant, τ is the time constant, $d\xi$ denotes the signal-dependent noise, and is given by

$$d\xi = \begin{bmatrix} c_1 & 0 \\ c_2 & 0 \end{bmatrix} \begin{bmatrix} u_x \\ u_y \end{bmatrix} d\eta_1 + \begin{bmatrix} 0 & c_2 \\ 0 & c_1 \end{bmatrix} \begin{bmatrix} u_x \\ u_y \end{bmatrix} d\eta_2 \tag{7.67}$$

TABLE 7.1   Parameters of the Linear Model

Parameters	Description	Value	Dimension
m	Hand mass	1.3	kg
b	Viscosity constant	10	N · s/m
τ	Time constant	0.05	s
c_1	Noise magnitude	0.075	
c_2	Noise magnitude	0.025	

where η_1 and η_2 are two standard and independent Brownian motions, $c_1 > 0$ and $c_2 > 0$ are constants describing the magnitude of the signal-dependent noise. The values of the parameters are specified in Table 7.1.

It is worth mentioning that this model is similar to many other linear models for describing arm movements; see, for example, [12, 17, 23, 45, 48, 50, 55], to which our ADP theory is also applicable.

7.3.2   Determining the Initial Stabilizing Control Policy

The proposed ADP-based online learning methodology requires an initial stabilizing control policy. To be more specific, we need to find an initial stabilizing feedback gain matrix $K_0 \in \mathbb{R}^{2\times6}$, such that the closed-loop matrix $A - BK_0$ is Hurwitz. By robust control theory [54], it is possible to find such a matrix K_0 if upper and lower bounds of the elements in both A and B are available and the pair (A, B) is stabilizable. Indeed, these bounds can be estimated by the CNS during the first several trials in the null-field (NF) immediately after the subject was seated.

For example, in the absence of disturbances (i.e., $f \equiv 0$), we have

$$A - BK_0 = \begin{bmatrix} 0 & I_2 & 0 \\ 0 & -\frac{b}{m}I_2 & \frac{1}{m}I_2 \\ 0 & 0 & -\frac{1}{\tau}I_2 \end{bmatrix} - \begin{bmatrix} 0 \\ 0 \\ \frac{1}{\tau}I_2 \end{bmatrix} K_0 \qquad (7.68)$$

Then, the first several trials in the NF can be interpreted as the exploration of an initial stabilizing feedback gain matrix K_0, by estimating the bounds on the parameters b, m, and τ, and solving a robust control problem. Indeed, if the CNS finds out that $b \in [-8, 12]$, $m \in [1, 1.5]$, and $\tau \in [0.03, 0.07]$ through the first several trials, the feedback gain matrix K_0 can thus be selected as

$$K_0 = \begin{bmatrix} 100 & 0 & 10 & 0 & 10 & 0 \\ 0 & 100 & 0 & 10 & 0 & 10 \end{bmatrix}. \qquad (7.69)$$

Once K_0 is obtained, the CNS can use the proposed ADP method to approximate the optimal control policy.

7.3.3 Selection of the Weighting Matrices

Here we model the selection of Q and R by the CNS. To begin with, notice that the symmetric weighting matrices $Q \in \mathbb{R}^{6\times6}$ and $R \in \mathbb{R}^{2\times2}$ contain as many as 24 independent parameters. To reduce redundancy, we assume

$$Q = \begin{bmatrix} Q_0 & 0 & 0 \\ 0 & 0.01Q_0 & 0 \\ 0 & 0 & 0.00005Q_0 \end{bmatrix}, \tag{7.70}$$

$$R = I_2 \tag{7.71}$$

where $Q_0 \in \mathbb{R}^{2\times2}$ is a symmetrical matrix.

Now, we assume that the CNS specifies the task-dependent matrix Q_0 as follows

$$Q_0(d_m, \theta) = \begin{bmatrix} \alpha_0 & 0 \\ 0 & \alpha_1 \end{bmatrix} + \alpha_2 d_m \begin{bmatrix} \cos\theta \\ \sin\theta \end{bmatrix} \begin{bmatrix} \cos\theta \\ \sin\theta \end{bmatrix}^T \tag{7.72}$$

where d_m denotes the magnitude of the largest deviation of the hand position in the previous trials from the unperturbed trajectory and θ is the angular position of the largest deviation (see Figure 7.2). Notice that both d_m and θ can be approximated by the CNS from previous trials. Positive constants α_0, α_1, and α_2 are determined based on long-term experiences of the subjects and are assumed to be invariant in our simulations.

The model (7.72) provides a unified way for the CNS to specify the weighting matrices with respect to different tasks. In the next two subsections, we show that this model is compatible with experimental observations.

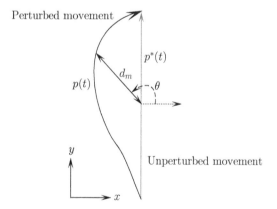

FIGURE 7.2   Illustration of the parameter θ and d_m that determine the weighting matrix Q. $p(t)$ and $p^*(t)$ are the perturbed and unperturbed movement trajectories, respectively. d_m denotes the magnitude of the largest deviation of the hand position from the unperturbed position. The angular position of this deviation with respect to the x-axis is denoted by θ. Notice that d_m and θ need to be changed only when the CNS realizes that the force field has been changed. *Source*: Jiang, 2014. Reproduced with permission of Springer.

7.3.4 Sensorimotor Control in a Velocity-Dependent Force Field

We used the proposed ADP method to simulate the experiment conducted by [13]. In that experiment, human subjects were seated and asked to move a parallel-link direct drive airmagnet floating manipulandum (PFM) to perform a series of forward arm reaching movements in the horizontal plane. All the subjects performed reaching movements from a start point located 0.25 m away from the target. The experiment tested human muscle stiffness and motor adaptation in a velocity-dependent force field (VF). The VF produced a stable interaction with the arm. The force exerted on the hand by the robotic interface in the VF was set to be

$$\begin{bmatrix} f_x \\ f_y \end{bmatrix} = \chi \begin{bmatrix} 13 & -18 \\ 18 & 13 \end{bmatrix} \begin{bmatrix} v_x \\ v_y \end{bmatrix} \tag{7.73}$$

where $\chi \in [2/3, 1]$ is a constant that can be adjusted to the subject's strength. In our simulation, we set $\chi = 0.7$.

Subjects in the experiment [13] first practiced in the NF. Trials were considered successful if they ended inside the target within the prescribed time 0.6 ± 0.1 s. After enough successful trials were completed, the force field was activated without notifying the subjects. Then, the subjects practiced in the VF until enough successful trials were achieved. After a short break, the subjects then performed several movements in the NF. These trials were called after-effects and were recorded to confirm that adaptation to the force field did occur. More details of the experimental setting can be found in [13, 16].

First, we simulated the movements in the NF. The simulation started with an initial stabilizing control policy that can be found by the CNS as explained in Section 7.3.2. During each trial, we collected the online data to update the control policy once. After enough trials, an approximate optimal control policy in the NF can be obtained. Also, the stiffness, which is defined as graphical depiction of the elastic restoring force corresponding to the unit displacement of the hand for the subject in the force fields [6], can be numerically computed. In addition, it can be represented in terms of an ellipse by plotting the elastic force produced by a unit displacement [39]. We ran the simulation multiple times with different values of α_0 and α_1. Then, we found that, by setting $\alpha_0 = 5 \times 10^4$ and $\alpha_1 = 1 \times 10^5$, the resultant stiffness ellipse has good consistency with experimental observations [13]. Since in the NF we have $d_m \approx 0$, the parameter α_2 does not contribute to the specification of Q_0.

Then, we proceeded with the simulations in the VF. The first trial was under the same control policy as used in the NF. After the first trial in the VF, the CNS can find the largest deviation of the first trial from the unperturbed movement. Hence, d_m and θ can be determined. Then, the CNS can modify the weighting matrix Q using the model (7.72). In our simulation, d_m and θ were set to be 0.15 and 144°. Then, the same Q matrix was unchanged for the rest trials in our simulation, since the CNS can soon realize that the new movements were conducted in the same force field. The ADP-based learning mechanism was applied starting from the second trial. The initial feedback gain matrix we used for the second trial was obtained by tripling the gains of the control policy in the NF. This is to model the experimental observation by [14], that muscle activities increased dramatically after the first trial. We ran the simulate multiple times with different values of α_2, and found good consistency with experimental results [13] by setting $\alpha_2 = 1 \times 10^6$. The stiffness ellipses are shown in Figure 7.5. One can compare it with the experimental observations in [13].

After 30 trials, the feedback control gain was updated to

$$K_{30} = \begin{bmatrix} 355.52 & 30.31 & 89.83 & -24.25 & 1.67 & -0.24 \\ -198.07 & 322.00 & -5.27 & 95.26 & -0.24 & 1.60 \end{bmatrix}$$

For comparison, the optimal feedback gain matrix is provided as follows:

$$K^* = \begin{bmatrix} 358.32 & 30.89 & 90.17 & -24.15 & 1.67 & -0.24 \\ -200.92 & 324.49 & -5.65 & 95.63 & -0.24 & 1.60 \end{bmatrix}$$

The simulated movement trajectories, the velocity curves, and the endpoint force curves are shown in Figures 7.3 and 7.4. It can be seen that the simulated movement trajectories in the NF are approximately straight lines, and the velocity curves along the y-axis are bell-shaped curves. These simulation results are consistent with experimental observations as well as the curves produced by the previous models [12, 38, 47]. After the subject was exposed to the VF, the first trial was simulated with the same feedback control policy as in the NF. This is because subjects in the experiment were not notified when the external force was activated. Apparently, this control policy was not optimal because the system dynamics was changed and the cost function was also different. Then, the ADP algorithm proceeded. In Figure 7.3, we see the first trial gives a movement trajectory which deviated far away from the straight path but eventually reached the target. Motor adaptation can be observed by comparing the first five consecutive trials. After 30 trials, the movement trajectories return to be straight lines, and the velocity curves become bell-shaped again. It implies that after 30 trials in the VF, the CNS can learn well the optimal control policy using real-time data, without knowing or using the precise system parameters. Finally, our numerical study shows clearly the after-effects of the subject behavior when the VF was suddenly deactivated.

To better illustrate the learning behavior in the VF, we define the movement time t_f of each trial as the time duration from the beginning of the trial until the hand-path enters and remains in the target area. Then, the movement times and distance were calculated and are shown in Figure 7.6.

7.3.5 Sensorimotor Control in a Divergent Field

Now, let us describe how we simulated the sensorimotor control system in a divergent field (DF) using the proposed ADP theory. In the experiment conducted by [6], the

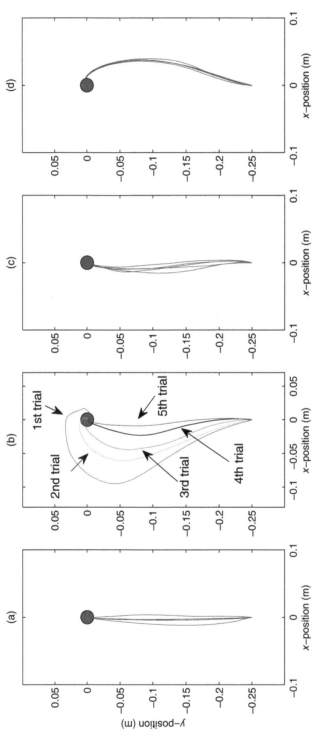

FIGURE 7.3 Simulated movement trajectories using the proposed learning scheme. a. Five successful movement trajectories of one subject in the NF. b. The first five consecutive movement trajectories of the subject when exposed to the VF. c. Five consecutive movement trajectories of the subject in the VF after 30 trials. d. Five independent after-effect trials.

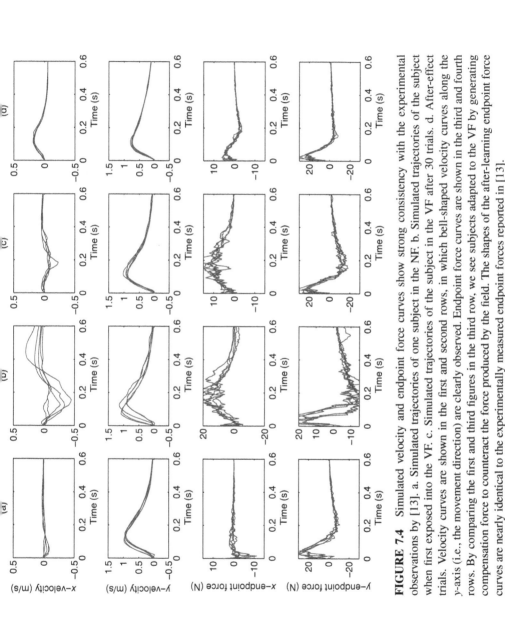

FIGURE 7.4 Simulated velocity and endpoint force curves show strong consistency with the experimental observations by [13]. a. Simulated trajectories of one subject in the NF. b. Simulated trajectories of the subject when first exposed into the VF. c. Simulated trajectories of the subject in the VF after 30 trials. d. After-effect trials. Velocity curves are shown in the first and second rows, in which bell-shaped velocity curves along the y-axis (i.e., the movement direction) are clearly observed. Endpoint force curves are shown in the third and fourth rows. By comparing the first and third figures in the third row, we see subjects adapted to the VF by generating compensation force to counteract the force produced by the field. The shapes of the after-learning endpoint force curves are nearly identical to the experimentally measured endpoint forces reported in [13].

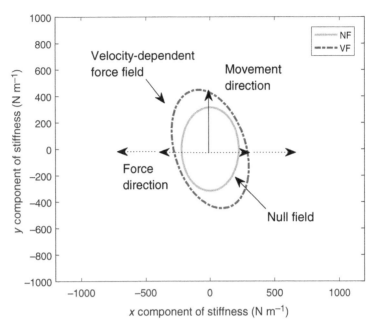

FIGURE 7.5   Illustration of the stiffness geometry to the VF. The stiffness in the VF increased significantly in the direction of the largest trajectory deviation (see Figure 7.2).

DF produced a negative elastic force perpendicular to the target direction, and was computed as

$$f = \begin{bmatrix} \beta & 0 \\ 0 & 0 \end{bmatrix} \begin{bmatrix} p_x \\ 0 \end{bmatrix} \tag{7.74}$$

where $\beta > 0$ is a sufficiently large constant such that the overall system is unstable. In our simulations, we set $\beta = 150$.

The simulation of the movements before the DF was applied is identical to the one described in Section 7.3.4, and an approximate optimal control policy in the NF has been obtained. However, this control policy is not stabilizing in the DF, and therefore an initial stabilizing control policy in the DF is needed. To be more specific, we need a matrix $K_0 \in \mathbb{R}^{2 \times 6}$ such that

$$A - BK_0 = \begin{bmatrix} 0 & 0 & 1 & 0 & 0 & 0 \\ 0 & 0 & 0 & 1 & 0 & 0 \\ \frac{\beta}{m} & 0 & -\frac{b}{m} & 0 & \frac{1}{m} & 0 \\ 0 & 0 & 0 & -\frac{b}{m} & 0 & \frac{1}{m} \\ 0 & 0 & 0 & 0 & -\frac{1}{\tau} & 0 \\ 0 & 0 & 0 & 0 & 0 & -\frac{1}{\tau} \end{bmatrix} - \begin{bmatrix} 0 & 0 \\ 0 & 0 \\ 0 & 0 \\ 0 & 0 \\ \frac{1}{\tau} & 0 \\ 0 & \frac{1}{\tau} \end{bmatrix} K_0 \tag{7.75}$$

is Hurwitz.

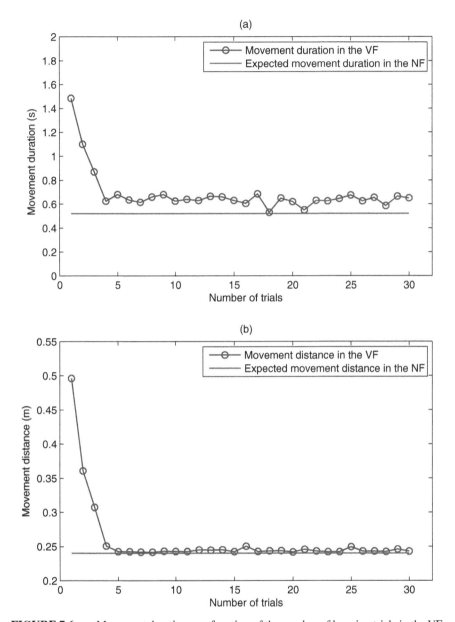

FIGURE 7.6   a. Movement duration as a function of the number of learning trials in the VF. b. Movement distance as a function of the number of learning trials in the VF. *Source*: Jiang, 2014. Reproduced with permission of Springer.

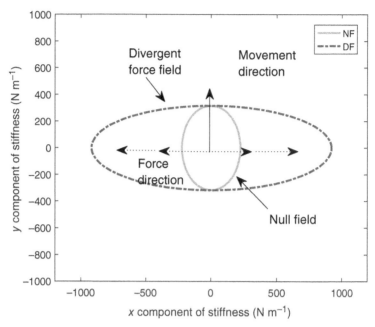

FIGURE 7.7 Illustration of stiffness geometry to the DF. The stiffness in the DF increased significantly in the direction of the external force, compared with the stiffness in the NF.

Therefore, we applied the same control policy learned in the NF to control the movements for the first five trials in the DF. As a result, unstable behaviors were observed in the first several trials (see Figure 7.8 b). After that, a stabilizing feedback control gain matrix K_0 became available to the CNS, since the CNS has estimated the range of the unknown parameters β, b, m, and τ, and should be able to find K_0 by solving a robust control problem. Here, we increased the first entry in the first row of the matrix $\hat{K}_{nf}$ by 300 and set the resultant matrix to be K_0, which is stabilizing. Then, we applied the proposed ADP algorithm with this K_0 as the initial stabilizing feedback gain matrix. Of course, K_0 can be selected in different ways. Some alternative models describing the learning process from instability to stability can also be found in [14, 46, 53, 55].

To obtain appropriate weighting matrices in the DF, we set $q_x = 1.5 \times 10^5$, $q_y = 10^5$, and $\theta = 15°$. This set of values can give good consistency between our simulation results and the experimental results [6]. Intuitively, we conjecture that the CNS increased the stiffness along the y-axis by assigning more weights to deal with the divergent force. Then, the stiffness ellipses can be numerically shown as in Figure 7.7. One can compare them with the experimental results reported by [6].

It can be seen in Figures 7.8 and 7.9 that the simulated movement trajectories in the NF are approximately straight lines and their velocity curves are bell-shaped. It is easy to notice that the movement trajectories differ slightly from trial to trial. This is due to the motor output variability caused by the signal-dependent noise. When

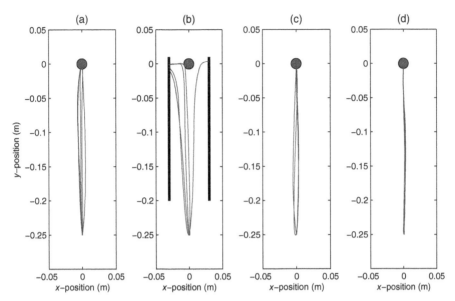

FIGURE 7.8 Simulated movement trajectories using the proposed learning scheme. a., Five successful movement trajectories of one subject in the NF. b. Five independent movement trajectories of the subject when exposed to the DF. The black lines on either side of trials in the DF indicate the safety zone, outside of which the force field was turned off. c. Five consecutive movement trajectories of the subject in the divergent field after 30 trials. d. Five consecutive after-effect trials.

the subject was first exposed to the DF, these variations were further amplified by the DF. As a result, unstable behaviors were observed in the first several trials.

In Figures 7.8 and 7.9, it is clear that, a stabilizing control policy is obtained, the proposed ADP scheme can be applied to generate an approximate optimal control policy. After 30 trials, the hand-path trajectories became approximately straight as in the NF. It implies that the subject has learned to adapt to the dynamics of the DF. Indeed, after 30 trials in the DF, the feedback gain matrix has been updated to

$$
K_{30} = \begin{bmatrix} 848.95 & 15.73 & 95.60 & 2.67 & 1.69 & 0.05 \\ 24.12 & 319.04 & 2.60 & 62.65 & 0.05 & 1.27 \end{bmatrix}
$$

For comparison purpose, the optimal feedback gain matrix for the ideal case with no noise is shown below.

$$
K_{df}^* = \begin{bmatrix} 853.67 & 15.96 & 96.07 & 2.70 & 1.70 & 0.05 \\ 24.39 & 321.08 & 2.63 & 62.86 & 0.05 & 1.27 \end{bmatrix}
$$

Finally, we simulated behavior of the subject when the force field is unexpectedly removed. From our simulation results, it is clear to see that the movement trajectories are even straighter than the trajectories in the NF. This is because the CNS has modified

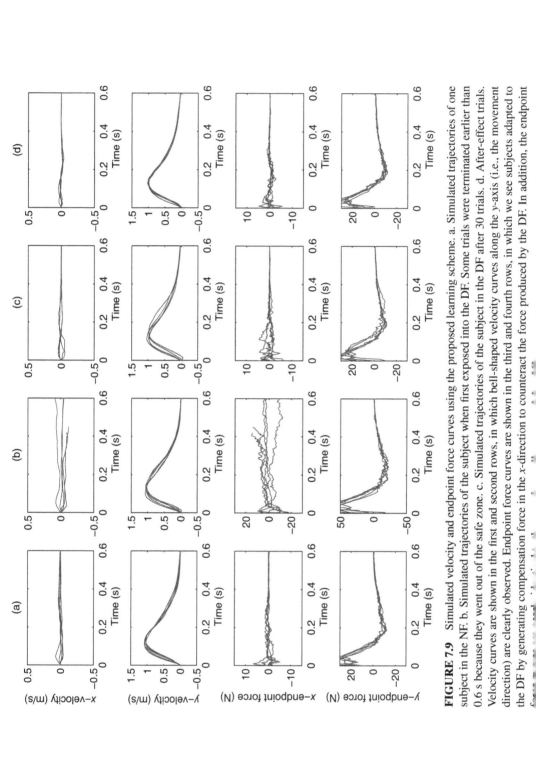

FIGURE 7.9 Simulated velocity and endpoint force curves using the proposed learning scheme. a. Simulated trajectories of one subject in the NF. b. Simulated trajectories of the subject when first exposed into the DF. Some trials were terminated earlier than 0.6 s because they went out of the safe zone. c. Simulated trajectories of the subject in the DF after 30 trials. d. After-effect trials. Velocity curves are shown in the first and second rows, in which bell-shaped velocity curves along the y-axis (i.e., the movement direction) are clearly observed. Endpoint force curves are shown in the third and fourth rows, in which we see subjects adapted to the DF by generating compensation force in the x-direction to counteract the force produced by the DF. In addition, the endpoint

the weighting matrices and put more weights on the displacement along the x-axis. As a result, the stiffness ellipses in the NF and DF are apparently different, because the stiffness increased significantly in the direction of the divergence force. The change of stiffness along the movement direction is not significant, as shown in our simulations. These characteristics match well with the experimental observations [6, 13].

Again, our simulation results show that the CNS can learn and find an approximate optimal control policy using real-time data, without knowing the precise system parameters.

7.3.6 Fitts's Law

According to [11], the movement duration t_f required to rapidly move to a target area is a function of the distance d to the target and the size of the target s, and a logarithmic law is formulated to represent the relationship among the three variables t_f, d, and s as follows

$$t_f = a + b \log_2 \left(\frac{d}{s} \right) \tag{7.76}$$

where a and b are two constants. Alternatively, the following power law is proposed in [41]:

$$t_f = a \left(\frac{d}{s} \right)^b \tag{7.77}$$

Here we validated our model by using both the log law and the power law. The target size s is defined as its diameter, and the distance is fixed as $d = 0.24$ m. We simulated the movement times from the trials in the NF, and the after-learning trials in the VF and DF. The data fitting results are shown in Figure 7.10 and Table 7.2. It can be seen that our simulation results are consistent with Fitts's law predictions.

7.4 NUMERICAL RESULTS: RADP-BASED SENSORIMOTOR CONTROL

In this section, we apply the proposed RADP algorithm to model arm movements in a VF. However, different from the previous section, we assume the mechanical device generating the forces was subject to certain time delay. Therefore, the dynamics of the mechanical device is treated as the dynamic uncertainty. We will also compare our simulation results with experimental observations [6, 13, 43]. The mathematical model for the motor system is the same as (7.64)–(7.66), with the parameters given in Table 7.1. We use the proposed RADP method to simulate the experiment conducted by [43]. We model the velocity-dependent force field using the following dynamic system

$$d \begin{bmatrix} f_x \\ f_y \end{bmatrix} = - \begin{bmatrix} \tau_f & 0 \\ 0 & \tau_f \end{bmatrix}^{-1} \times \left(\begin{bmatrix} f_x \\ f_y \end{bmatrix} - \begin{bmatrix} -10.1 & -11.2 \\ -11.2 & 11.1 \end{bmatrix} \begin{bmatrix} v_x \\ v_y \end{bmatrix} \right) dt \tag{7.78}$$

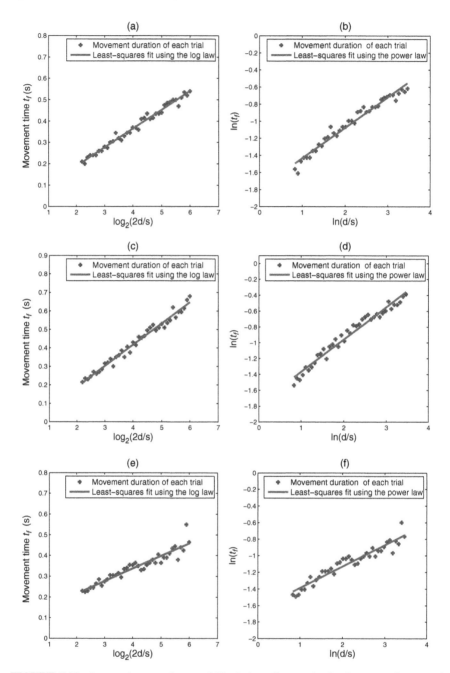

FIGURE 7.10 Log and power forms of Fitts's law. Crosses in the first row, the second row, and the third row represent movement times simulated in the NF, the VF, and the DF, respectively. Solid lines in a, b, c are least-squares fits using the log Fitts's law, and solid lines in b, d, f are the least-squares fits using the power Fitts's law.

TABLE 7.2   Data Fitting for the Log Law and Power Law

Parameters	NF	VF	DF
a (Log law)	0.0840	0.1137	0.0829
b (Log law)	0.0254	−0.0376	0.0197
a (Power law)	0.3401	0.4101	0.3468
b (Power law)	−1.7618	−1.7796	−1.8048

In the experiment [43], each subject was asked to move a cursor from the center of a workspace to a target at an angle randomly chosen from the set $\{0°, 45°, \ldots, 315°\}$, and at a distance of 0.1 m. After one target was reached, the next target, randomly selected, was presented.

There were four different stages in the experiment. First, the subject made arm movements without the force field. Second, the force field was applied without notifying the subject. Third, the subject adapted to the force field. Finally, the force field was suddenly removed and the after-learning effects were observed.

Before the force field was activated, the initial control policy we assumed was the same as the one in the previous simulation. Once the CNS noticed the velocity-dependent field, the new weighting matrices was replaced by

$$Q_1 = \begin{bmatrix} 10^4 & 0 & 0 & 0 \\ 0 & 1000 & 0 & 0 \\ 0 & 0 & 1 & 0 \\ 0 & 0 & 0 & 1 \end{bmatrix}, Q_2 = I_2,$$

$$R_1 = R_2 = \begin{bmatrix} 0.1 & 0 \\ 0 & 0.1 \end{bmatrix} \tag{7.79}$$

The movement performance in the four different stages were simulated using the proposed RADP algorithm and the results are given in Figure 7.11. In addition, we plotted the velocity curves of the arm movement during the first three stages as shown in Figure 7.12. Interestingly, consistency can be found by comparing the curves in Figure 7.12 with Figure 10 in [43].

7.5 DISCUSSION

7.5.1 Non-Model-Based Learning

Most of the previous models for sensorimotor control have concluded that the CNS knows precisely the knowledge of the motor system and its interacting environment [8, 12, 17, 27, 40, 42, 47, 48, 49]. The computation of optimal control laws is based on this assumption. By contrast, the proposed ADP methodology is a non-model-based approach and informs that the optimal control policy is derived using the real-time sensory data and is robust to dynamic uncertainties such as signal-dependent noise.

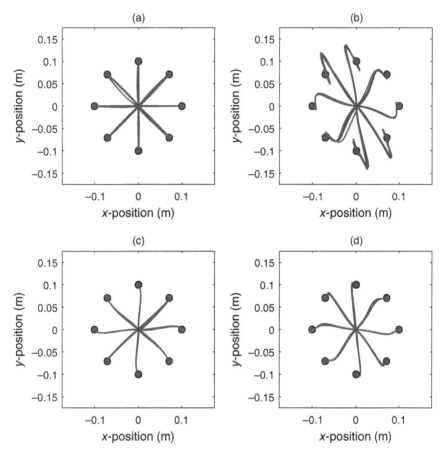

FIGURE 7.11 Simulation of hand trajectories in the VF using the proposed RADP-based learning scheme. Movements originate at the center. a. Simulation of hand trajectories in the null force field. b. Simulated Performance during initial exposure to the force field. c. Simulated hand trajectories in the force field after 30 trials. d. Simulated after-effects of adaptation to the force field.

In the absence of external disturbances, our model can generate typically observed position, velocity, and endpoint force curves as produced by the previous models. As one of the key differences with existing sensorimotor models, our proposed computational mechanism suggests that, when confronted with unknown environments and imprecise dynamics, the CNS may update and improve its command signals for movement through learning and repeated trials.

In the presence of perturbations, most of the previous models have relied on sensory prediction errors to form an estimate of the perturbation [2, 31, 52]. However, this viewpoint is difficult to be justified theoretically and has not been convincingly validated by experiments. Indeed, evidence against this source-identified adaptation is reported by [21], where a self-generated perturbation was created but it was not identified or used in formulating the control policy. This is consistent with the learning

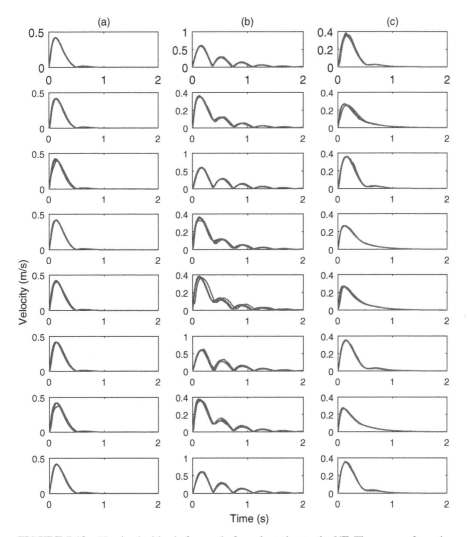

FIGURE 7.12 Hand velocities before and after adaptation to the VF. The curves, from the first row to the eighth row, are for targets at 0°, 45°,..., 315°. a. Hand velocities in the NF before exposure to force field. b. Hand velocities upon initial exposure to the VF. c. Hand velocities after adaptation to the VF.

scheme introduced in this chapter. Indeed, our ADP-based learning scheme does not identify the dynamics of the force fields. Instead, optimal control policies in the presence of force fields are directly obtained through successive approximations. By simulating the experiments conducted by [6] and [13], we have found that our computational results match well with the experimental results. In particular, our simulation results show gradual adaptation to the unknown force fields, with nearly identical movement trajectories in the first several consecutive trials reported in the experiment [13]. The simulated post-learning velocity and endpoint force curves fit well with the experimental observations [6, 13]. Our simulations clearly demonstrated the after-effects phenomenon.

7.5.2   Stability and Convergence Properties

Several reinforcement learning-based models for motor adaptation have been developed in the past literature [9, 24]. However, it is not easy to analyze the convergence and the stability properties of the learning schemes. In this chapter, we have extended the ADP theory to continuous-time linear systems with signal-dependent noise, and applied this theory to model sensorimotor control. An added value of the proposed ADP methodology is that rigorous convergence and stability analysis is given and is surprisingly simple by means of linear optimal control theory.

7.5.3   Muscle Stiffness

Several practical methods for experimentally measuring stiffness have been proposed in the past literature [5, 16, 18], and the changes of stiffness in force fields were reported by [6] and [13]. However, how the CNS modifies the stiffness geometry and achieves optimal motor behavior remains a largely open question. Burdet et al. [6] suggested that the CNS minimizes the hand-path error relative to a straight line joining the start position and the target center. This optimization problem does not involve any system dynamics, and cannot link the modification of stiffness to optimal feedback control theory. On the other hand, the stiffness may not be well analyzed using other models based on finite-horizon optimal control theory (see, e.g., [17, 47]). This is because those models use time-varying control policies, leading to continuous change of the stiffness during the movement.

   In the ADP-based model, time-invariant control policies are computed, and it is comparably easy to analyze the muscle stiffness by studying the position feedback gains. Our modeling methodology implies that the change of stiffness results from the modification of the weighing matrices by the CNS and the change of the system dynamics. Indeed, our simulation results provide similar stiffness ellipses as those measured in experiments [6, 13]. In addition, our model suggests that different stiffness geometries of different individuals may be a consequence of different weighting matrices they selected. Therefore, compared with other models of motor control and motor adaptation, our modeling strategy can explain naturally the change of stiffness observed in experiments from the viewpoint of optimal feedback control [35].

7.5.4   Data Fitting for the Weighting Matrices

The weighting matrices we used in the numerical simulation are selected such that our resultant computational results can have qualitative consistency with experimental results [6, 13]. If accurate human motor data become available, better fits for the weights can be obtained using a two-loop optimization approach [25]. The inner-loop uses the proposed ADP method to approximate an optimal control policy and generate the stiffness ellipses. The outer-loop compares the error between the simulated stiffness ellipses with experimental observations and adjusts the parameters q_x, q_y, and θ to minimize the error.

7.5.5   Infinite-Horizon Optimal Control

During each trial, a time-invariant control policy is suggested in our methodology. The time-invariant control policy has a main advantage that movement duration does not need to be pre-fixed by the CNS. This seems more realistic because the duration of each movement is different from each other due to the signal-dependent noise and is difficult to be pre-fixed. Todorov and Jordan [48] suggested that if the target is not reached at the predicted reaching time, the CNS can similarly plan an independent new trajectory between the actual position of the hand and the final target, and the final trajectory will be the superposition of all the trajectories. By contrast, our model matches the intuitive notion that the motor system keeps moving the hand toward the target until it is reached, and much less computational burden is required. Our simulation results match well with Fitts's law predictions. In addition, this type of control policies can also be used to analytically derive the Fitts's law as illustrated by [40].

7.5.6   Comparison with Iterative Learning Control

Iterative learning control (ILC) is an open-loop control scheme that iteratively learns a feedforward signal based on the tracking error of the previous trials [4]. It has also been used to model motor learning [55]. Compared with ILC, the ADP-based learning method has at least three advantages. First, conventional ILC method uses open-loop control policy for each individual trial. Hence, it may not explain the feedback control mechanism involved in each individual movement, which is essential in generating bell-shaped velocity-curves. Second, ILC assumes the external disturbance is iteration invariant. However, motor uncertainties do not satisfy this assumption, since no two movements are exactly the same. Third, in the ILC model, learning only happens among different trials. Therefore, it cannot model the online learning during a single trial. On the other hand, this online learning process can be modeled by our ADP method, since the ADP scheme gives the flexibility to specify the amount of time that is needed for collecting online data and updating the control policy. Indeed, if the time duration is set to be less than the time needed for one trial, then online learning during one movement can be simulated.

7.5.7   Connection Between Optimality and Robustness

Although optimal control theory is the dominant paradigm for understanding motor behavior, and optimization based models can explain many aspects of sensor motor control, it is not clear if the CNS keeps using the optimal control policy when dynamic uncertainty occurs. Experimental results obtained by [21] show that the control scheme employed by the CNS in the presence of external disturbances may only be suboptimal because the control scheme they observed experimentally is not energy efficient. Indeed, in the presence of dynamic uncertainties, guaranteeing optimality and stability becomes a non-trivial task.

The RADP studies the stability of interconnected systems by analyzing the gain conditions (7.32), (7.33), and (7.42)–(7.44), which are inspired by a simplified version of the nonlinear small-gain theorem [28]. As shown previously, these conditions can be satisfied by choosing suitable cost functions with appropriately designed weighting matrices [26]. It should be mentioned that the control policy generated from our RADP method is optimal for the nominal/reduced system and remains stable and retains suboptimality in the presence of dynamic uncertainty. The change of stiffness in the divergent force field was reported by [6]. However, they assumed that the stiffness was modified to minimize some cost function associated with the minimum-jerk model [12] which does not involve optimal feedback control theory. Alternatively, in the RADP theory, the change of stiffness can be interpreted as a direct result of the change of weighting matrices.

Hence, the proposed RADP theory is compatible with the optimal control theory and the experimental results observed in the past literature. More importantly, it provides a unified framework that naturally connects optimality and robustness to explain the motor behavior with/without uncertainties.

7.6 NOTES

From different aspects, many theories have been proposed to explain the computational nature of sensorimotor control; see the review article [15]. One widely accepted view is that the CNS prefers trajectories produced by minimizing some cost function. This perspective has inspired quite a few optimization-based models for motor control (see [8, 12, 17, 19, 27, 40, 42, 47–49], and references therein). These models can explain many characteristics of motor control, such as approximately straight movement trajectories and the bell-shaped velocity curves reported by [38]. However, these models assume the CNS knows and uses the dynamics of both the motor system and the interactive environment. Consequently, an indirect control scheme is assumed. Namely, the CNS first identifies all the system dynamics, and then finds the optimal control policies based on the identified information. This identification-based idea has also been used to study motor adaptation under external perturbations [2, 3, 7, 31, 43, 52]. Nevertheless, this viewpoint is difficult to be justified theoretically and has not been convincingly validated by experiments. Using self-generated perturbation, [21] reported that disturbance may not be identified by the CNS, and the control policy may not necessarily be optimal in the presence of uncertainties. Indeed, when uncertainties, especially dynamic uncertainties, occur, it becomes difficult to maintain not only optimality but also stability. Since the optimization-based model may not be suitable to study the behavior and stability of motor systems, developing a new theoretical modeling framework is not only necessary but also of great importance.

There are two main advantages of the modeling strategy introduced in this chapter. First of all, as a non-model-based approach, RADP shares some essential features with reinforcement learning (RL) [44], which is originally inspired by learning mechanisms observed in biological systems. RL concerns how an agent should modify its actions to interact with the unknown environment and to achieve a long-term goal. In addition, certain brain areas that can realize the steps of RL have been discussed by [10]. Like in many other ADP-based methods, the proposed RADP theory solves the Bellman equation [1] iteratively using real-time sensory information and can avoid the *curse of dimensionality* in conventional dynamic programming. Also, rigorous convergence analysis can be performed. Second, instead of identifying the dynamics of the overall dynamic system, we decompose the system into an interconnection of a nominal/reduced system with measurable state variables and the dynamic uncertainty (or, unmodeled dynamics) with unmeasurable state variables and unknown system order. Then, we design the robust optimal control policy for the overall system using partial state-feedback. In this way, we can preserve optimality for the nominal deterministic model as well as guarantee robust stability for the overall system. Compared with identification-based models, this modeling strategy is more realistic for sensorimotor systems for at least two reasons. First, identifying the exact model of both the motor system and the uncertain environment is not an easy task. Second, it is time-consuming and would yield slow response if the sensorimotor system first estimates all system variables before taking actions.

Detailed learning algorithms are presented in this chapter, and numerical studies are also provided. Interestingly, our computational results match well with experiments reported from the past literature [6, 13, 43]. The proposed theory also provides a unified theoretical framework that connects optimality and robustness. In addition, it links the stiffness geometry to the selection of the weighting matrices in the cost function. Therefore, we argue that the CNS may use RADP-like learning strategy to coordinate movements and to achieve successful adaptation in the presence of static and/or dynamic uncertainties. In the absence of the dynamic uncertainties, the learning strategy reduces to an ADP-like mechanism.

In summary, in this chapter we developed ADP and RADP methods for linear stochastic systems with signal-dependent noise with an objective to model goal-oriented sensorimotor control systems. An appealing feature of this new computational mechanism of sensorimotor control is that the CNS does not rely upon the a priori knowledge of systems dynamics and the environment to generate a command signal for hand movement. Our theory can explain the change of stiffness geometry from a perspective of adaptive optimal feedback control versus nonadaptive optimal control theory [40, 48, 47]. In particular, the RADP theory not only gives computational results which are compatible with experimental data [6, 13, 43], but also provides a unified framework to study robustness and optimality simultaneously. We therefore argue that the human motor system may use ADP and RADP-like mechanisms to control movements.

REFERENCES

[1] R. E. Bellman. *Dynamic Programming*. Princeton University Press, Princeton, NJ, 1957.

[2] M. Berniker and K. Kording. Estimating the sources of motor errors for adaptation and generalization. *Nature Neuroscience*, 11(12):1454–1461, 2008.

[3] N. Bhushan and R. Shadmehr. Computational nature of human adaptive control during learning of reaching movements in force fields. *Biological Cybernetics*, 81(1):39–60, 1999.

[4] D. A. Bristow, M. Tharayil, and A. G. Alleyne. A survey of iterative learning control. *IEEE Control Systems Magazine*, 26(3):96–114, 2006.

[5] E. Burdet, R. Osu, D. Franklin, T. Yoshioka, T. Milner, and M. Kawato. A method for measuring endpoint stiffness during multi-joint arm movements. *Journal of Biomechanics*, 33(12):1705–1709, 2000.

[6] E. Burdet, R. Osu, D. W. Franklin, T. E. Milner, and M. Kawato. The central nervous system stabilizes unstable dynamics by learning optimal impedance. *Nature*, 414(6862):446–449, 2001.

[7] P. R. Davidson and D. M. Wolpert. Motor learning and prediction in a variable environment. *Current Opinion in Neurobiology*, 13(2):232–237, 2003.

[8] J. Diedrichsen, R. Shadmehr, and R. B. Ivry. The coordination of movement: optimal feedback control and beyond. *Trends in Cognitive Sciences*, 14(1):31–39, 2010.

[9] K. Doya. Reinforcement learning in continuous time and space. *Neural Computation*, 12(1):219–245, 2000.

[10] K. Doya, H. Kimura, and M. Kawato. Neural mechanisms of learning and control. *IEEE Control Systems Magazine*, 21(4):42–54, 2001.

[11] P. M. Fitts. The information capacity of the human motor system in controlling the amplitude of movement. *Journal of Experimental Psychology*, 47(6):381–391, 1954.

[12] T. Flash and N. Hogan. The coordination of arm movements: An experimentally confirmed mathematical model. *The Journal of Neuroscience*, 5(7):1688–1703, 1985.

[13] D. W. Franklin, E. Burdet, R. Osu, M. Kawato, and T. E. Milner. Functional significance of stiffness in adaptation of multijoint arm movements to stable and unstable dynamics. *Experimental Brain Research*, 151(2):145–157, 2003.

[14] D. W. Franklin, E. Burdet, K. P. Tee, R. Osu, C.-M. Chew, T. E. Milner, and M. Kawato. CNS learns stable, accurate, and efficient movements using a simple algorithm. *The Journal of Neuroscience*, 28(44):11165–11173, 2008.

[15] D. W. Franklin and D. M. Wolpert. Computational mechanisms of sensorimotor control. *Neuron*, 72(3):425–442, 2011.

[16] H. Gomi and M. Kawato. Equilibrium-point control hypothesis examined by measured arm stiffness during multijoint movement. *Science*, 272:117–120, 1996.

[17] C. M. Harris and D. M. Wolpert. Signal-dependent noise determines motor planning. *Nature*, 394:780–784, 1998.

[18] N. Hogan. The mechanics of multi-joint posture and movement control. *Biological Cybernetics*, 52(5):315–331, 1985.

[19] N. Hogan and T. Flash. Moving gracefully: Quantitative theories of motor coordination. *Trends in Neurosciences*, 10(4):170–174, 1987.

[20] R. A. Horn and C. R. Johnson. *Matrix Analysis*. Cambridge University Press, 1990.

[21] T. E. Hudson and M. S. Landy. Adaptation to sensory-motor reflex perturbations is blind to the source of errors. *Journal of Vision*, 12(1):1–10, 2012.

[22] K. Itô. Stochastic integral. *Proceedings of the Japan Academy, Series A, Mathematical Sciences*, 20(8):519–524, 1944.

[23] J. Izawa, T. Rane, O. Donchin, and R. Shadmehr. Motor adaptation as a process of reoptimization. *The Journal of Neuroscience*, 28(11):2883–2891, 2008.

[24] J. Izawa and R. Shadmehr. Learning from sensory and reward prediction errors during motor adaptation. *PLoS Computational Biology*, 7(3):e1002012, 2011.

[25] Y. Jiang, S. Chemudupati, J. M. Jorgensen, Z. P. Jiang, and C. S. Peskin. Optimal control mechanism involving the human kidney. In: Proceedings of the 50th IEEE Conference on Decision and Control and European Control Conference (CDC-ECC), pp. 3688–3693, Orlando, FL, 2011.

[26] Y. Jiang and Z. P. Jiang. Robust adaptive dynamic programming. In: D. Liu and F. Lewis, editors, *Reinforcement Learning and Adaptive Dynamic Programming for Feedback Control*, Chapter 13, pp. 281–302. John Wiley & Sons, 2012.

[27] Y. Jiang, Z. P. Jiang, and N. Qian. Optimal control mechanisms in human arm reaching movements. In: Proceedings of the 30th Chinese Control Conference, pp. 1377–1382, Yantai, China, 2011.

[28] Z. P. Jiang, A. R. Teel, and L. Praly. Small-gain theorem for ISS systems and applications. *Mathematics of Control, Signals and Systems*, 7(2):95–120, 1994.

[29] D. Kleinman. On an iterative technique for Riccati equation computations. *IEEE Transactions on Automatic Control*, 13(1):114–115, 1968.

[30] D. Kleinman. On the stability of linear stochastic systems. *IEEE Transactions on Automatic Control*, 14(4):429–430, 1969.

[31] K. P. Kording, J. B. Tenenbaum, and R. Shadmehr. The dynamics of memory as a consequence of optimal adaptation to a changing body. *Nature Neuroscience*, 10(6):779–786, 2007.

[32] M. Krstic and H. Deng. *Stabilization of Nonlinear Uncertain Systems*. Springer, 1998.

[33] M. Krstic and Z.-H. Li. Inverse optimal design of input-to-state stabilizing nonlinear controllers. *IEEE Transactions on Automatic Control*, 43(3):336–350, 1998.

[34] H. J. Kushner. *Stochastic Stability*. Springer, Berlin, Heidelberg, 1972.

[35] F. Lewis, D. Vrabie, and V. Syrmos. *Optimal Control*, 3rd ed. John Wiley & Sons, Inc., Hoboken, NJ, 2012.

[36] D. Liu and E. Todorov. Evidence for the flexible sensorimotor strategies predicted by optimal feedback control. *The Journal of Neuroscience*, 27(35):9354–9368, 2007.

[37] L. Ljung. *System Identification*. John Wiley & Sons, 1999.

[38] P. Morasso. Spatial control of arm movements. *Experimental Brain Research*, 42(2):223–227, 1981.

[39] F. A. Mussa-Ivaldi, N. Hogan, and E. Bizzi. Neural, mechanical, and geometric factors subserving arm posture in humans. *The Journal of Neuroscience*, 5(10):2732–2743, 1985.

[40] N. Qian, Y. Jiang, Z. P. Jiang, and P. Mazzoni. Movement duration, Fitts's law, and an infinite-horizon optimal feedback control model for biological motor systems. *Neural Computation*, 25(3):697–724, 2013.

[41] R. A. Schmidt and T. D. Lee. *Motor Control and Learning: A Behavioral Emphasis*, 5th ed. Human Kinetics, 2011.

[42] S. H. Scott. Optimal feedback control and the neural basis of volitional motor control. *Nature Reviews Neuroscience*, 5(7):532–546, 2004.

[43] R. Shadmehr and F. A. Mussa-Ivaldi. Adaptive representation of dynamics during learning of a motor task. *The Journal of Neuroscience*, 14(5):3208–3224, 1994.

[44] R. S. Sutton and A. G. Barto. *Reinforcement Learning: An Introduction.* Cambridge University Press, 1998.

[45] H. Tanaka, J. W. Krakauer, and N. Qian. An optimization principle for determining movement duration. *Journal of Neurophysiology*, 95(6):3875–3886, 2006.

[46] K. P. Tee, D. W. Franklin, M. Kawato, T. E. Milner, and E. Burdet. Concurrent adaptation of force and impedance in the redundant muscle system. *Biological Cybernetics*, 102(1):31–44, 2010.

[47] E. Todorov. Stochastic optimal control and estimation methods adapted to the noise characteristics of the sensorimotor system. *Neural Computation*, 17(5):1084–1108, 2005.

[48] E. Todorov and M. I. Jordan. Optimal feedback control as a theory of motor coordination. *Nature Neuroscience*, 5(11):1226–1235, 2002.

[49] Y. Uno, M. Kawato, and R. Suzuki. Formation and control of optimal trajectory in human multijoint arm movement: Minimum torque-change model. *Biological Cybernetics*, 61(2):89–101, 1989.

[50] K. Wei and K. Körding. Uncertainty of feedback and state estimation determines the speed of motor adaptation. *Frontiers in Computational Neuroscience*, 4:11, 2010.

[51] J. L. Willems and J. C. Willems. Feedback stabilizability for stochastic systems with state and control dependent noise. *Automatica*, 12(3):277–283, 1976.

[52] D. M. Wolpert and Z. Ghahramani. Computational principles of movement neuroscience. *Nature Neuroscience*, 3:1212–1217, 2000.

[53] C. Yang, G. Ganesh, S. Haddadin, S. Parusel, A. Albu-Schaeffer, and E. Burdet. Humanlike adaptation of force and impedance in stable and unstable interactions. *IEEE Transactions on Robotics*, 27(5):918–930, 2011.

[54] K. Zhou, J. C. Doyle, and K. Glover. *Robust and Optimal Control.* Prentice Hall, New Jersey, 1996.

[55] S.-H. Zhou, D. Oetomo, Y. Tan, E. Burdet, and I. Mareels. Modeling individual human motor behavior through model reference iterative learning control. *IEEE Transactions on Biomedical Engineering*, 59(7):1892–1901, 2012.

APPENDIX A

BASIC CONCEPTS IN NONLINEAR SYSTEMS

Here we review some important tools from modern nonlinear control; see, for instance, [4, 5, 7, 9, 11, 16], and references therein for the details. See [8] for more recent developments in nonlinear systems and control.

A.1 LYAPUNOV STABILITY

Let us start with the following autonomous nonlinear system

$$\dot{x} = f(x) \tag{A.1}$$

where $f : D \subset \mathbb{R}^n \to \mathbb{R}^n$ is a locally Lipschtiz function satisfying $f(0) = 0$.

Theorem A.1.1 ([9]) *Consider system (A.1). Let $V(x)$ be a continuously differentiable function defined in a domain $D \subset \mathbb{R}^n$ containing the origin. If*

(1) $V(0) = 0$,
(2) $V(x) > 0, \forall x \in D/\{0\}$, *and*
(3) $\dot{V}(x) \leq 0, \forall x \in D$, *then the origin of system (A.1) is stable. Moreover, if $\dot{V}(x) < 0, \forall x \in D/\{0\}$, then the origin is asymptotically stable.*

Robust Adaptive Dynamic Programming, First Edition. Yu Jiang and Zhong-Ping Jiang.
© 2017 by The Institute of Electrical and Electronics Engineers, Inc. Published 2017 by John Wiley & Sons, Inc.

Furthermore, if $V(x) > 0$, $\forall x \neq 0$, and

$$|x| \to \infty \Rightarrow V(x) \to \infty \tag{A.2}$$

and $\dot{V}(x) < 0$, $\forall x \neq 0$, then the origin is globally asymptotically stable.

Theorem A.1.2 (LaSalle's Theorem [9]) *Let $f(x)$ be a locally Lipschitz function defined over a domain $D \subset \mathbb{R}^n$, and $\Omega \subset \mathbb{R}^n$ be a compact set that is positively invariant with respect to system (A.1). Let $V(x)$ be a continuously differentiable function defined over D such that $\dot{V} \leq 0$ in Ω. Let E be the set of all points in Ω where $\dot{V}(x) = 0$, and M be the largest invariant set in E. Then every solution starting in Ω approaches M as $t \to \infty$.*

A.2 ISS AND THE SMALL-GAIN THEOREM

Consider the system

$$\dot{x} = f(x, u) \tag{A.3}$$

where $x \in \mathbb{R}^n$ is the state, $u \in \mathbb{R}^m$ is the input, and $f : \mathbb{R}^n \times \mathbb{R}^m \to \mathbb{R}^n$ is locally Lipschitz.

Definition A.2.1 ([14, 15]) *The system (A.3) is said to be* input-to-state stable *(ISS) with gain γ if, for any measurable essentially bounded input u and any initial condition $x(0)$, the solution $x(t)$ exists for every $t \geq 0$ and satisfies*

$$|x(t)| \leq \beta(|x(0)|, t) + \gamma(\|u\|) \tag{A.4}$$

where β is of class $\mathcal{KL}$ and γ is of class $\mathcal{K}$.

Definition A.2.2 ([17]) *A continuously differentiable function V is said to be an ISS-Lyapunov function for the system (A.3) if V is positive definite and radially unbounded, and satisfies the following implication*

$$|x| \geq \chi(|u|) \Rightarrow \nabla V(x)^T f(x, u) \leq -\kappa(|x|) \tag{A.5}$$

where κ is positive definite and χ is of class $\mathcal{K}$.

Next, consider an interconnected system described by

$$\dot{x}_1 = f_1(x_1, x_2, v), \tag{A.6}$$
$$\dot{x}_2 = f_2(x_1, x_2, v) \tag{A.7}$$

where, for $i = 1, 2$, $x_i \in \mathbb{R}^{n_i}$, $v \in \mathbb{R}^{n_v}$, $f_i : \mathbb{R}^{n_1} \times \mathbb{R}^{n_2} \times \mathbb{R}^{n_v} \to \mathbb{R}^{n_i}$ is locally Lipschitz.

Assumption A.2.3   *For each $i = 1, 2$, there exists an ISS-Lyapunov function V_i for the x_i subsystem such that the following hold:*

1. *there exist functions $\underline{\alpha}_i, \bar{\alpha}_i \in \mathcal{K}_\infty$, such that*

$$\underline{\alpha}_i(|x_i|) \leq V_i(x_i) \leq \bar{\alpha}_i(|x_i|), \forall x_i \in \mathbb{R}^{n_i}, \tag{A.8}$$

2. *there exist functions $\chi_i, \gamma_i \in \mathcal{K}$ and $\alpha_i \in \mathcal{K}_\infty$, such that*

$$\nabla V_1(x_1)^T f_1(x_1, x_2, v) \leq -\alpha_1(V_1(x_1)), \tag{A.9}$$

if $V_1(x_1) \geq \max\{\chi_1(V_2(x_2)), \gamma_1(|v|)\}$, and

$$\nabla V_2(x_2)^T f_2(x_1, x_2, v) \leq -\alpha_2(V_2(x_2)), \tag{A.10}$$

if $V_2(x_2) \geq \max\{\chi_2(V_1(x_1)), \gamma_2(|v|)\}$.

Based on the ISS-Lyapunov functions, the following theorem gives the small-gain condition, under which the ISS property of the interconnected system can be achieved.

Theorem A.2.4 ([16])   *Under Assumption A.2.3, if the following small-gain condition holds:*

$$\chi_1 \circ \chi_2(s) < s, \forall s > 0, \tag{A.11}$$

then, the interconnected system (A.6) and (A.7) is ISS with respect to v as the input.

Under Assumption A.2.3 and the small-gain condition (A.11), let $\hat{\chi}_1$ be a function of class $\mathcal{K}_\infty$ such that

1. $\hat{\chi}_1(s) \leq \chi_1^{-1}(s), \forall s \in [0, \lim_{s \to \infty} \chi_1(s))$,
2. $\chi_2(s) \leq \hat{\chi}_1(s), \forall s \geq 0$.

Then, as shown in [6], there exists a class $\mathcal{K}_\infty$ function $\sigma(s)$ which is continuously differentiable over $(0, \infty)$ and satisfies $\frac{d\sigma}{ds}(s) > 0$ and $\chi_2(s) < \sigma(s) < \hat{\chi}_1(s), \forall s > 0$.
In [6], it is also shown that the function

$$V_{12}(x_1, x_2) = \max\{\sigma(V_1(x_1)), V_2(x_2)\} \tag{A.12}$$

is positive definite and radially unbounded. In addition,

$$\dot{V}_{12}(x_1, x_2) < 0 \tag{A.13}$$

holds almost everywhere in the state space, whenever

$$V_{12}(x_1, x_2) \geq \eta(|v|) > 0 \tag{A.14}$$

for some class $\mathcal{K}_\infty$ function η.

APPENDIX B

SEMIDEFINITE PROGRAMMING AND SUM-OF-SQUARES PROGRAMMING

B.1 SDP AND SOSP

As a subfield of convex optimization, semidefinite programming (SDP) is a broad generalization of linear programming. It concerns with the optimization of a linear function subject to linear matrix inequality constraints, and is of great theoretic and practical interest [1].

A standard SDP problem can be formulated as the following problem of minimizing a linear function of a variable $y \in \mathbb{R}^{n_0}$ subject to a linear matrix inequality.

Problem B.1.1 (Semidefinite Programming [18])

$$\min_{y} \quad c^T y \tag{B.1}$$

$$F_0 + \sum_{i=1}^{n_0} y_i F_i \geq 0 \tag{B.2}$$

where $c \in \mathbb{R}^{n_0}$ is a constant column vector, and $F_0, F_1, \dots, F_{n_0} \in \mathbb{R}^{m_0 \times m_0}$ are $n_0 + 1$ symmetric constant matrices. Notice that (B.2) is not an element-wise inequality, but a positive semidefinite constraint.

SDPs can be solved using several commercial or non-commercial software packages, such as the MATLAB-based solver CVX [2].

Robust Adaptive Dynamic Programming, First Edition. Yu Jiang and Zhong-Ping Jiang.
© 2017 by The Institute of Electrical and Electronics Engineers, Inc. Published 2017 by John Wiley & Sons, Inc.

Definition B.1.2 (Sum of Squares [1])   *A polynomial $p(x)$ is an SOS if there exist polynomials $q_1, q_2, \ldots, q_{m_0}$ such that*

$$p(x) = \sum_{i=1}^{m} q_i^2(x) \tag{B.3}$$

Remark B.1.3   *In Section 4.5, the SOS property has been further extended to a non-negative condition for a general class of non-polynomial nonlinear functions.*

An SOS program is a convex optimization problem of the following form.

Problem B.1.4 (SOS Program [1])

$$\min_{y} \quad b^T y \tag{B.4}$$

$$\text{s.t.} \quad p_i(x; y) \text{ are SOS}, \ i = 1, 2, \ldots, k_0 \tag{B.5}$$

where $p_i(x; y) = a_{i0}(x) + \sum_{j=1}^{n_0} a_{ij}(x) y_j$, and $a_{ij}(x)$ are given polynomials.

In [1, p. 74], it has been pointed out that SOS programs are in fact equivalent to SDPs [1, 18]. The conversion from an SOS to an SDP can be performed either manually or automatically using, for example, SOSTOOLS [12, 13], YALMIP [10], and GloptiPoly [3].

APPENDIX C

PROOFS

C.1 PROOF OF THEOREM 3.1.4

Before proving Theorem 3.1.4, let us first give the following lemma.

Lemma C.1.1 *Consider the conventional policy iteration algorithm described in (3.9) and (3.10). Suppose u_i is a globally stabilizing control policy and there exists $V_i(x) \in C^1$ with $V(0) = 0$, such that (3.9) holds. Let u_{i+1} be defined as in (3.10). Then, under Assumption 3.1.2, the followings are true.*

(1) $V_i(x) \geq V^*(x)$;

(2) *for any $V_{i-1} \in P$, such that*

$$\nabla V_{i-1}[(f(x) + g(x)u_i] + q(x) + u_i^T R(x)u_i \leq 0, \tag{C.1}$$

we have $V_i \leq V_{i-1}$;

(3) $\nabla V_i^T(f(x) + g(x)u_{i+1}) + q(x) + u_{i+1}^T R(x)u_{i+1} \leq 0.$

Robust Adaptive Dynamic Programming, First Edition. Yu Jiang and Zhong-Ping Jiang.
© 2017 by The Institute of Electrical and Electronics Engineers, Inc. Published 2017 by John Wiley & Sons, Inc.

Proof:

(1) Under Assumption 3.1.2, we have

$$
\begin{aligned}
0 &= \nabla V_i^T(x)(f(x) + g(x)u_i) + r(x, u_i) \\
&\quad - \nabla V^{*T}(x)(f(x) + g(x)u^*) - r(x, u^*) \\
&= (\nabla V_i - \nabla V^*)^T(f(x) + g(x)u_i) + r(x, u_i) \\
&\quad - (\nabla V^*)^T g(x)(u^* - u_i) - r(x, u^*) \\
&= (\nabla V_i - \nabla V^*)^T(f(x) + g(x)u_i) \\
&\quad + (u_i - u^*)^T R(x)(u_i - u^*)
\end{aligned}
$$

Therefore, for any $x_0 \in \mathbb{R}^n$, along the trajectories of system (3.1) with $u = u_i$ and $x(0) = x_0$, we have

$$
\begin{aligned}
& V_i(x_0) - V^*(x_0) \\
&= \int_0^T (u_i - u^*)^T R(u_i - u^*)dt + V_i(x(T)) - V^*(x(T)) \quad \text{(C.2)}
\end{aligned}
$$

Since u_i is globally stabilizing, we know $\lim_{T \to +\infty} V_i(x(T)) = 0$ and $\lim_{T \to +\infty} V^*(x(T)) = 0$. Hence, letting $T \to +\infty$, from (C.2) it follows that $V_i(x_0) \geq V^*(x_0)$. Since x_0 is arbitrarily selected, we have $V_i(x) \geq V^*(x), \forall x \in \mathbb{R}^n$.

(2) Let $q_i(x) \geq 0$, such that

$$
\nabla V_i^T(x)(f(x) + g(x)u_i) + r(x, u_i) = -q_i(x), \forall x \in \mathbb{R}^n. \quad \text{(C.3)}
$$

Then,

$$
(\nabla V - \nabla V_i)^T(f + gu_i) + q_i(x) = 0. \quad \text{(C.4)}
$$

Similar as in (1), along the trajectories of system (3.1) with $u = u_i$ and $x(0) = x_0$, we can show

$$
V(x_0) - V_i(x_0) = \int_0^\infty q_i(x)dt \quad \text{(C.5)}
$$

Hence, $V(x) \leq V_i(x), \forall x \in \mathbb{R}^n$.

(3) By definition,

$$
\begin{aligned}
& \nabla V_i^T(f(x) + g(x)u_{i+1}) + q(x) + u_{i+1}^T R(x)u_{i+1} \\
&= \nabla V_i^T(f(x) + g(x)u_i) + q(x) + u_i^T R(x)u_i \\
&\quad + \nabla V_i^T g(x)(u_{i+1} - u_i) + u_{i+1}^T R(x)u_{i+1} - u_i^T R(x)u_i
\end{aligned}
$$

$$= -2u_{i+1}^T R(x)(u_{i+1} - u_i) + u_{i+1}^T R(x)u_{i+1} - u_i^T R(x)u_i$$
$$= +2u_{i+1}^T R(x)u_i - u_{i+1}^T R(x)u_{i+1} - u_i^T R(x)u_i$$
$$\leq 0 \tag{C.6}$$

The proof is complete. ∎

Proof of Theorem 3.1.4

We first prove (1) and (2) by induction. To be more specific, we will show that (1) and (2) are true and $V_i \in \mathcal{P}$, for all $i = 0, 1, \ldots$

(i) If $i = 1$, by Assumption 3.1.1 and Lemma C.1.1 (1), we immediately know (1) and (2) hold. In addition, by Assumptions 3.1.1 and 3.1.2, we have $V^* \in \mathcal{P}$ and $V_0 \in \mathcal{P}$. Therefore, $V_i \in \mathcal{P}$.

(ii) Suppose (1) and (2) hold for $i = j > 1$ and $V_j \in \mathcal{P}$. We show that (1) and (2) also hold for $i = j + 1$ and $V_{j+1} \in \mathcal{P}$.

Indeed, since $V^* \in \mathcal{P}$ and $V_j \in \mathcal{P}$, by the induction assumption, we know $V_{j+1} \in \mathcal{P}$.

Next, by Lemma C.1.1 (3), we have

$$\nabla V_{j+1}[f(x) + g(x)u_{j+2}] + q(x) + u_{j+2}^T R(x)u_{j+2} \leq 0 \tag{C.7}$$

As a result, along the solutions of system (3.1) with $u = u_{j+2}$, we have

$$\dot{V}_{j+1}(x) \leq -q(x) \tag{C.8}$$

Notice that, since $V_{j+1} \in \mathcal{P}$, it is a well-defined Lyapunov function for the closed-loop system (3.1) with $u = u_{j+2}$. Therefore, u_{j+2} is globally stabilizing, that is, (2) holds for $i = j + 1$.

Then, by Lemma C.1.1 (2), we have

$$V_{j+2} \leq V_{j+1} \tag{C.9}$$

Together with the induction Assumption, it follows that

$$V^* \leq V_{j+2} \leq V_{j+1} \tag{C.10}$$

Hence, (1) holds for $i = j + 1$.

Now, let us prove (3). If such a pair (V, u) exists, we immediately know

$$u = -\frac{1}{2}R^{-1}g^T \nabla V \tag{C.11}$$

Hence, V is the solution to the HJB in (3.3). Also,

$$V^* \leq V \leq V_0 \Rightarrow V^* \in \mathcal{P} \tag{C.12}$$

However, as discussed in Section 3.1, solution to the HJB equation (3.3) must be unique. As a result, $V^* = V$ and $u^* = u$.

The proof is complete.

C.2   PROOF OF THEOREM 3.2.3

To begin with, given $\hat{u}_i$, let $\tilde{V}_i(x)$ be the solution of the following equation with $\tilde{V}_i(0) = 0$

$$\nabla \tilde{V}_i(x)^T (f(x) + g(x)\hat{u}_i(x)) + q(x) + \hat{u}_i^T R(x)\hat{u}_i = 0 \tag{C.13}$$

and denote $\tilde{u}_{i+1}(x) = -\frac{1}{2}R^{-1}(x)g^T(x)\nabla \tilde{V}_i(x)^T$.

Lemma C.2.1   *For each $i \geq 0$, we have* $\lim\limits_{N_1,N_2 \to \infty} \hat{V}_i(x) = \tilde{V}_i(x)$, $\lim\limits_{N_1,N_2 \to \infty} \hat{u}_{i+1}(x) = \tilde{u}_{i+1}(x)$, $\forall x \in \Omega$.

Proof: By definition

$$\tilde{V}_i(x(t_{k+1})) - \tilde{V}_i(x(t_k))$$
$$= -\int_{t_k}^{t_{k+1}} [q(x) + \hat{u}_i^T R(x)\hat{u}_i + 2\tilde{u}_{i+1}^T R(x)\hat{v}_i]dt \tag{C.14}$$

Let $\tilde{c}_{i,j}$ and $\tilde{w}_{i,j}$ be the constant weights such that $\tilde{V}_i(x) = \sum_{j=1}^{\infty} \tilde{c}_{i,j}\phi_j(x)$ and $\tilde{u}_{i+1}(x) = \sum_{j=1}^{\infty} \tilde{w}_{i,j}\psi_j(x)$. Then, by (3.17) and (C.14), we have $e_{i,k} = \theta_{i,k}^T \bar{W}_i + \xi_{i,k}$, where

$$\bar{W}_i = \left[\tilde{c}_{i,1}\ \tilde{c}_{i,2}\ \cdots\ \tilde{c}_{i,N_1}\ \tilde{w}_{i,1}\ \tilde{w}_{i,2}\ \cdots\ \tilde{w}_{i,N_2} \right]^T$$
$$- \left[\hat{c}_{i,1}\ \hat{c}_{i,2}\ \cdots\ \hat{c}_{i,N_1}\ \hat{w}_{i,1}\ \hat{w}_{i,2}\ \cdots\ \hat{w}_{i,N_2} \right]^T,$$

$$\xi_{i,k} = \sum_{j=N_1+1}^{\infty} \tilde{c}_{i,j} \left[\phi_j(x(t_{k+1})) - \phi_j(x(t_k)) \right]$$
$$+ \sum_{j=N_2+1}^{\infty} \tilde{w}_{i,j} \int_{t_k}^{t_{k+1}} 2\psi_j^T R(x)\hat{v}_i dt.$$

Since the weights are found using the least-squares method, we have

$$\sum_{k=1}^{l} e_{i,k}^2 \le \sum_{k=1}^{l} \xi_{i,k}^2$$

Also, notice that,

$$\sum_{k=1}^{l} \bar{W}_i^T \theta_{i,k}^T \theta_{i,k} \bar{W}_i = \sum_{k=1}^{l} (e_{i,k} - \xi_{i,k})^2$$

Then, under Assumption 3.2.2, it follows that

$$|\bar{W}_i|^2 \le \frac{4}{\delta} \max_{1 \le k \le l} \xi_{i,k}^2.$$

Therefore, given any arbitrary $\epsilon > 0$, we can find $N_{10} > 0$ and $N_{20} > 0$, such that when $N_1 > N_{10}$ and $N_2 > N_{20}$, we have

$$|\hat{V}_i(x) - \tilde{V}_i(x)|$$
$$\le \sum_{j=1}^{N_1} |c_{i,j} - \hat{c}_{i,j}||\phi_j(x)| + \sum_{j=N_1+1}^{\infty} |c_{i,j}\phi_j(x)|$$
$$\le \frac{\epsilon}{2} + \frac{\epsilon}{2} = \epsilon, \forall x \in \Omega. \tag{C.15}$$

Similarly, $|\hat{u}_{i+1}(x) - \tilde{u}_{i+1}(x)| \le \epsilon$. The proof is complete.   ∎

We now prove Theorem 3.2.3 by induction.

(1) If $i = 0$ we have $\tilde{V}_0(x) = V_0(x)$, and $\tilde{u}_1(x) = u_1(x)$. Hence, the convergence can directly be proved by Lemma A.1.

(2) Suppose for some $i > 0$, we have $\lim_{N_1,N_2 \to \infty} \hat{V}_{i-1}(x) = V_{i-1}(x)$, $\lim_{N_1,N_2 \to \infty} \hat{u}_i(x) = u_i(x)$, $\forall x \in \Omega$. By definition, we have

$$|V_i(x(t)) - \tilde{V}_i(x(t))|$$
$$= |\int_t^{\infty} [\hat{u}_i^T(x)R(x)\hat{u}_i(x) - u_i^T(x)R(x)u_i(x)] \, dt|$$
$$+ 2|\int_t^{\infty} u_{i+1}^T(x)R(x)[\hat{u}_i(x) - u_i(x)] \, dt|$$
$$+ 2|\int_t^{\infty} [\tilde{u}_{i+1}(x) - u_{i+1}(x)]^T R(x)\hat{v}_i dt|, \forall x \in \Omega.$$

By the induction assumptions, we know

$$0 = \lim_{N_1,N_2\to\infty} \int_t^\infty \left[\hat{u}_i^T(x)R(x)\hat{u}_i(x) - u_i^T(x)R(x)u_i(x)\right] dt \tag{C.16}$$

$$0 = \lim_{N_1,N_2\to\infty} \int_t^\infty u_{i+1}^T(x)R(x)\left[\hat{u}_i(x) - u_i(x)\right] dt \tag{C.17}$$

Also, by Assumption 3.2.2, we conclude

$$\lim_{N_1,N_2\to\infty} |u_{i+1}(x) - \tilde{u}_{i+1}(x)| = 0 \tag{C.18}$$

and

$$\lim_{N_1,N_2\to\infty} |V_i(x) - \tilde{V}_i(x)| = 0 \tag{C.19}$$

Finally, since

$$|\hat{V}_i(x) - V_i(x)| \le |V_i(x) - \tilde{V}_i(x)| + |\tilde{V}_i(x) - \hat{V}_i(x)|$$

and by the induction assumption, we have

$$\lim_{N_1,N_2\to\infty} |V_i(x) - \hat{V}_i(x)| = 0 \tag{C.20}$$

Similarly, we can show

$$\lim_{N_1,N_2\to\infty} |u_{i+1}(x) - \hat{u}_i(x)| = 0 \tag{C.21}$$

The proof is thus complete.

REFERENCES

[1] G. Blekherman, P. A. Parrilo, and R. R. Thomas (editors). *Semidefinite Optimization and Convex Algebraic Geometry.* SIAM, Philadelphia, PA, 2013.

[2] M. Grant and S. Boyd. CVX: MATLAB software for disciplined convex programming, version 2.1. Available at http://cvxr.com/cvx. Accessed December 2013.

[3] D. Henrion and J.-B. Lasserre. GloptiPoly: Global optimization over polynomials with Matlab and SeDuMi. *ACM Transactions on Mathematical Software*, 29(2):165–194, 2003.

[4] A. Isidori. *Nonlinear Control Systems*, Vol. 2. Springer, 1999.

[5] Z. P. Jiang and I. Mareels. A small-gain control method for nonlinear cascaded systems with dynamic uncertainties. *IEEE Transactions on Automatic Control*, 42(3):292–308, 1997.

[6] Z. P. Jiang, I. M. Mareels, and Y. Wang. A Lyapunov formulation of the nonlinear small-gain theorem for interconnected ISS systems. *Automatica*, 32(8):1211–1215, 1996.

[7] Z. P. Jiang, A. R. Teel, and L. Praly. Small-gain theorem for ISS systems and applications. *Mathematics of Control, Signals and Systems*, 7(2):95–120, 1994.

[8] I. Karafyllis and Z. P. Jiang. *Stability and Stabilization of Nonlinear Systems*. Springer, 2011.

[9] H. K. Khalil. *Nonlinear Systems*, 3rd ed. Prentice Hall, Upper Saddle River, NJ, 2002.

[10] J. Lofberg. YALMIP: A toolbox for modeling and optimization in Matlab. In: Proceedings of 2004 IEEE International Symposium on Computer Aided Control Systems Design, pp. 284–289, Las Vegas, Nevada, 2004.

[11] R. Marino and P. Tomei. *Nonlinear Control Design: Geometric, Adaptive and Robust*. Prentice Hall, 1996.

[12] A. Papachristodoulou, J. Anderson, G. Valmorbida, S. Prajna, P. Seiler, and P. A. Parrilo. SOSTOOLS: Sum of squares optimization toolbox for MATLAB, 2013. Available at http://www.eng.ox.ac.uk/control/sostools, http://www.cds.caltech.edu/sostools, and http://www.mit.edu/ parrilo/sostools/

[13] S. Prajna, A. Papachristodoulou, and P. A. Parrilo. Introducing SOSTOOLS: A general purpose sum of squares programming solver. In: Proceedings of the 41st IEEE Conference on Decision and Control, pp. 741–746, Las Vegas, Nevada, 2002.

[14] E. D. Sontag. Smooth stabilization implies coprime factorization. *IEEE Transactions on Automatic Control*, 34(4):435–443, 1989.

[15] E. D. Sontag. Further facts about input to state stabilization. *IEEE Transactions on Automatic Control*, 35(4):473–476, 1990.

[16] E. D. Sontag. Input to state stability: Basic concepts and results. In: *Nonlinear and Optimal Control Theory*, pp. 163–220. Springer, 2008.

[17] E. D. Sontag and Y. Wang. On characterizations of the input-to-state stability property. *Systems & Control Letters*, 24(5):351–359, 1995.

[18] L. Vandenberghe and S. Boyd. Semidefinite programming. *SIAM Review*, 38(1):49–95, 1996.

INDEX

action, 1–6
adaptive dynamic programming, *see* ADP
adaptive optimal control, 29–30
admissible, 50, 55
  control policy, 36, 39, 67
ADP, 2
affine nonlinear system, 5, 35, 46
after-effects, 156–157, 168, 170
agent, 1–2, 4–5, 173
airmagnet floating manipulandum, 156
algebraic Riccati equation, *see* ARE
algorithm
  generalized SOS-based policy iteration, 68
  large-scale RADP, 123
  linear off-policy ADP, 22
  linear on-policy ADP, 18–19
  linear policy iteration, *see* Kleinman's algorithm
  linear stochastic on-policy ADP, 142
  linear stochastic on-policy RADP, 153
  linear two-phase policy iteration, 150
  nonlinear conventional policy iteration, 37, 51

nonlinear off-policy ADP, 42
nonlinear off-policy RADP, 94
non-polynomial GADP, 70
off-policy GRADP, 102
polynomial GADP, 63
SOS-based policy iteration, 55–56
SOS-based robust policy iteration, 101
AlphaGo, 2
approximate dynamic programming, *see* ADP
ARE, 3, 11–12, 29, 88–89, 92, 94, 105, 115, 122, 125, 139, 144, 147
arm movement, 153–154, 165, 167

Barto, 1
Bellman, 2, 81, 173
bell-shaped, 157, 159, 162, 164, 171–172
Brownian motion, 138, 143, 145, 154

car suspension system, 43, 70–71
central nervous system, *see* CNS
CNS, 6, 137, 151, 154–157, 162–163, 165, 167–168, 170–173
compact set, 39, 41–43, 52, 103

Robust Adaptive Dynamic Programming, First Edition. Yu Jiang and Zhong-Ping Jiang.
© 2017 by The Institute of Electrical and Electronics Engineers, Inc. Published 2017 by John Wiley & Sons, Inc.

IEEE PRESS SERIES ON SYSTEMS SCIENCE AND ENGINEERING

Editor:
MengChu Zhou, *New Jersey Institute of Technology and Tongji University*

Co-Editors:
Han-Xiong Li, *City University of Hong-Kong*
Margot Weijnen, *Delft University of Technology*

The focus of this series is to introduce the advances in theory and applications of systems science and engineering to industrial practitioners, researchers, and students. This series seeks to foster system-of-systems multidisciplinary theory and tools to satisfy the needs of the industrial and academic areas to model, analyze, design, optimize and operate increasingly complex man-made systems ranging from control systems, computer systems, discrete event systems, information systems, networked systems, production systems, robotic systems, service systems, and transportation systems to Internet, sensor networks, smart grid, social network, sustainable infrastructure, and systems biology.